Study Guide
MCAT
PHYSICS

Joseph Boone

PRENTICE HALL, Upper Saddle River, NJ 07458

Executive Editor: Alison Reeves
Production Editor: Kim Dellas
Supplement Cover Designer: Liz Nemeth
Special Projects Manager: Barbara A. Murray
Supplement Cover Manager: Paul Gourhan
Manufacturing Buyer: Ben Smith
Assistant Editor: Wendy Rivers

Printed in the United States of America

10 9 8 7 6 5

ISBN 0-13-627951-1

Prentice-Hall International (UK) Limited,London
Prentice-Hall of Australia Pty. Limited, Sydney
Prentice-Hall Canada Inc., Toronto
Prentice-Hall Hispanoamericana, S.A., Mexico
Prentice-Hall of India Private Limited, New Delhi
Prentice-Hall of Japan, Inc., Tokyo
Pearson Education Asia Pte. Ltd., Singapore
Editora Prentice-Hall do Brasil, Ltda., Rio de Janeiro

Introduction

Physics is not about memorizing equations and trying to decide which equation to use for the particular problem at hand. Physics is about understanding a few basic physical principles and using them to analyze problems. Reasoning skills are much more important than an ability to memorize.

To talk the language of physics, you must know the vocabulary. This means you must know the definitions of words that are commonly used in physics. Some of these words are instantaneous velocity, displacement, force, power, torque, and electric charge. Many definitions are written in equation form, and these definitions must be thoroughly understood. In addition to the basic definitions, there are certain relationships that must also be learned in order to do certain problems, but the number of equations that you need to memorize is not large. Do not fall into the habit of trying to memorize too many equations. Memorizing equations is not a substitute for understanding the physical principles.

Most of the physics questions on the MCAT require more thought and reasoning than simply plugging in numbers and turning the crank. Often incorrect answers on a multiple choice test can be eliminated by checking the units, or by checking to see if the answer is in the right ball park.

This study guide is not intended to teach you physics. You should have passed physics with a respectable grade and learned the basic principles and definitions. This guide is designed to refresh your memory about the topics you should have covered, and to provide review questions to help orient you to the type of questions you are likely to find on the MCAT.

Contents

1 Units and Vectors

1.1 Units and Measurements

Physics is a study of the rules that govern how various objects behave in the physical universe. The most important elements of physics are the principles that enable us to predict the behavior of physical objects. These principles are often stated in equation form. Examples include the law of gravity, Newton's laws of motion, and the principles of electricity and magnetism.

The use of these physical principles to predict the outcome of an experiment involves making measurements and performing numerical calculations. These numbers usually have units associated with them. All physical quantities can be described in terms of a small number of **fundamental or base units**. Quantities that are expressed as combinations of base units are called **derived units**. For example, speed is a derived unit that is a unit of length divided by a unit of time. Acceleration is a derived unit that is a velocity divided by a time, or a unit of length divided by the square of a unit of time. The system of units used by the scientific community is the metric system, which is now officially called the **International System of Units** (abbreviated **SI**).

The important fundamental or base SI units are listed below.

Quantity	Unit	Abbreviation
Length	meter	m
Time	second	s
Mass	kilogram	kg
Electric current	ampere	A
Temperature	kelvin	K
Amount of substance	mole	mol
Luminous intensity	candela	cd

Some important derived SI units are listed below.

Quantity	Unit	Abbreviation and equivalent	
Force	newton	N	($1\ \text{N} = 1\ \text{kg·m/s}^2$)
Energy and Work	joule	J	($1\ \text{J} = 1\ \text{N·m} = 1\ \text{kg·m}^2/\text{s}^2$)
Power	watt	W	($1\ \text{W} = 1\ \text{J/s} = 1\ \text{kg·m}^2/\text{s}^3$)
Pressure	pascal	Pa	($1\ \text{Pa} = 1\ \text{N/m}^2 = 1\ \text{kg·m/s}^2$)
Frequency	hertz	Hz	($1\ \text{Hz} = 1\ \text{cycle/s}$)
Electric charge	coulomb	C	($1\ \text{C} = 1\ \text{A·s}$)
Electric potential	volt	V	($1\ \text{V} = 1\ \text{J/C}$)

1

Problem Solving Strategy: The correct answer to a problem must have the appropriate units. In multiple choice tests, a quick check of the units of the answers can sometimes eliminate one or more of the choices. The fundamental quantities used in physical descriptions are called dimensions. Analyzing a problem to see that the answer has the correct units or checking an equation to make sure the units are consistent is called **dimensional analysis**.

Example 1.1

On a level surface, a bullet is fired with a velocity of v at an angle of θ above the horizontal. If g is the acceleration of gravity, the time of flight of the bullet is given by the formula:

A. $2vg \sin \theta$

B. $\dfrac{2g}{v} \sin \theta$

C. $\dfrac{2v^2}{g} \sin \theta$

D. $\dfrac{2v}{g} \sin \theta$

The answer has units of time so we need to check the units of the choices. We will use [L] for length and [T] for time. (Use [M] for mass in problems where it is needed.) The acceleration of gravity has units of $[L/T^2]$ and the velocity has units of $[L/T]$.

A. vg has units of $[L/T][L/T^2] = [L^2/T^3]$

B. g/v has units of $[L/T^2]/[L/T] = 1/[T]$

C. v^2/g has units of $[L/T]^2/[L/T^2] = [L^2/T^2]/[L/T^2] = [L]$

D. v/g (the reciprocal of B) has units of $[L/T]/[L/T^2] = [T]$

The only answer with the correct units is D.

Problem Solving Strategy: When calculating an answer by multiplying and/or dividing various given quantities, the units after the manipulations must agree with the units of the quantity you were asked to find.

Example 1.2

If the numbers given in a problem are 12 m and 3 s and the question asks for a velocity or speed, the quantities must be manipulated so that the answer has the correct units. Speed is measured in m/s, therefore, the answer could not be obtained by multiplying the two quantities together since this process gives the wrong units. (12 m)(3 s) = 36 m·s However, the answer could be (12 m)/(3 s) = 4 m/s since this division gives the correct units. The units can be checked in this way, but *not* the numerical value. For example, if 12 m is the diameter of a circular path and 3 s is the time to move around the circle, the speed would be: $\pi(12 \text{ m})/(3 \text{ s}) = 12.6 \text{ m/s}$

1.2 Vectors and Scalars

In physics, we often encounter quantities that have an obvious direction in space. Most of these directed quantities are called vectors. A quantity with a direction is a vector if it *also* obeys the law of vector addition.

A **vector** is fully defined if its magnitude (including the appropriate units) and its direction are given. For example, a velocity of 30 m/s north is a vector since its magnitude (30 m/s) and its direction (north) are specified. Other vector quantities include accelerations, forces, and torques.

Quantities that do not have a direction associated with them are called **scalars**. Examples include temperature, time, work, energy, pressure, and speed (speed is the *magnitude* of a velocity and hence does not have a direction). On typed pages, vectors are represented by bold letters such as:

A, B, C, a, b, c

Since vectors have a magnitude (like 30 m/s) and a direction (like north), they are usually represented graphically by arrows. The arrow points in the direction associated with the vector, and the length of the arrow is drawn proportional to the magnitude (a vector of 30 m/s would be twice as long as a vector of 15 m/s).

Two vectors are defined to be equal if they have the same magnitude (length) and point in the same direction. Vectors do *not* need to be located at the same position in space to be equal. A vector may be transported to a different location and its value will not change if its length and direction are not altered.

It should be obvious that two different types of vectors (vectors with different units) cannot be compared in any meaningful way. For example, a vector with a magnitude of 30 m/s and a vector with a magnitude of 15 m/s^2 cannot be compared since one is a velocity vector and the other is an acceleration vector (30 m/s is not twice as long as 15 m/s^2). We cannot compare apples and oranges.

Vector Addition

Vectors **A** and **B** can be added (if they are the same type) to produce the vector sum (written **A** + **B**) by proceeding as follows. Transport vector **B** so its tail touches the head of vector **A**. The sum (vector **A** + **B**) is the vector drawn from the tail of vector **A** to the head of vector **B**. The sum of two vectors **A** and **B** is illustrated graphically in the figure.

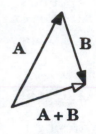

For convenience let us rename vector (**A** + **B**) and call it vector **C**. Symbolically we could now write:

$$\mathbf{A} + \mathbf{B} = \mathbf{C}$$

This equation states that vector **C** is equivalent to vector **A** plus vector **B**. Therefore, vector **C** could be replaced by the two vectors **A** and **B**, much like the scalar number 8 can be replaced by the two numbers 3 and 5.

It is easy to add two vectors graphically, but it may be extremely tedious to actually calculate the magnitude and direction of the vector **C** = **A** + **B**. Even if you are told that **A** has a magnitude of 4 m/s and **B** has a magnitude of 3 m/s, you do not have enough information to find **A** + **B**. Unless the two vectors are parallel and point in the same direction, their sum is not 7 m/s.

Resolution of a Vector into Components

We have seen that calculating the sum of two vectors is generally a very tedious process. However, if the two vectors are parallel or even perpendicular, then addition becomes a fairly simple process. Consider the perpendicular vectors **A** and **B** shown at the right.

The sum of these two vectors is the vector drawn from the tail of **A** to the head of **B**. Since the two vectors are perpendicular, we can find the *magnitude* of their sum by applying the Pythagorean theorem:

$$|\mathbf{A} + \mathbf{B}| = \sqrt{3^2 + 4^2}\ \text{m/s} = 5\ \text{m/s}$$

Working with vectors is easier if all the vectors are either parallel, or perpendicular. This can be accomplished by introducing a coordinate system and then replacing each vector that does not lie along one of the axes with two vectors lying along the x and y axes whose sum is equal to the original vector. This process is called **resolving a vector into its components**. Resolving vector **A** into its components is the process of finding the two vectors that lie along the coordinate axes and have a sum equal to **A**. We will call these two vectors $\mathbf{A}_x$ and $\mathbf{A}_y$.

$\mathbf{A}_x$ and $\mathbf{A}_y$ are vectors, but since they lie along either the x axis or the y axis, we can indicate their direction with a plus or minus sign. We also have to remember which axis they lie along, but that will not be too hard unless we get careless with the subscripts. For convenience, we will write the components of vector **A** as the scalars A_x and A_y. These scalar components will be either positive or negative quantities depending on their directions. The process is illustrated below.

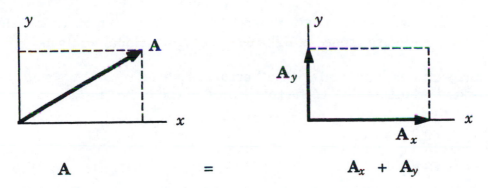

$$\mathbf{A} = \mathbf{A}_x + \mathbf{A}_y$$

A **unit vector** is a vector with a magnitude of *one* (it has no units). Any vector can be written as the product of a scalar (like 23 m/s) and a unit vector. The unit vector is used to indicate the direction. Three important unit vectors point along the coordinate axes. Unit vectors pointing in the positive x, y, and z directions are represented by the symbols **i**, **j**, and **k** respectively. As an illustration, the vector (**A** = 23 m/s **i**) is a velocity vector with a magnitude of 23 m/s that points in the *positive x* direction. The vector (**B** = -23 m/s **j**) has a magnitude of 23 m/s but points in the *negative y* direction.

Example 1.3

A car is traveling along the x axis at a speed of 30 m/s in the negative x direction. Write the velocity vector in unit vector notation.

The velocity vector is: **v** = - (30 m/s) **i**

Problem Solving Strategy: The only vectors that can be manipulated easily (without a calculator) are those whose components are the legs of a 3-4-5 triangle, the legs of a 30°-60° right triangle, or the legs of a 45° right triangle. You should know the sine, cosine, and tangent for the angles of 0°, 30°, 45°, 60°, 90°, and for the angles of the 3-4-5 triangle.

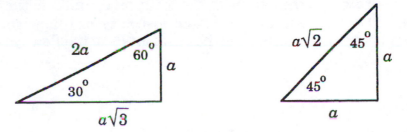

Figure 1.1 A 30°-60° right triangle and a 45° right triangle. The ratio of the sides should be memorized. For example: sin 30°=1/2; sin 45°=1/√2; etc.

5

Questions and Problems

1. If the frequency of blue light is 6×10^{14} cycles/second and its wavelength is 5×10^{-7} m, find the speed of light by using dimensional analysis.

2. A projectile is launched with a horizontal speed of 100 m/s and a vertical speed of 30 m/s. What is the total speed of the projectile?
   A. 97 m/s
   B. 104 m/s
   C. 115 m/s
   D. 130 m/s

3. On a level surface, a bullet is fired at an angle of θ above the horizontal with a speed of v. The horizontal distance covered by the bullet is given by the formula: (Hint: Use dimensional analysis.)
   A. $vg \sin 2\theta$
   B. $\dfrac{g}{v^2} \sin 2\theta$
   C. $\dfrac{v}{g} \sin 2\theta$
   D. $\dfrac{2v^2}{g} \sin\theta \, \cos\theta$

4. Some complicated fluid flow meter is governed by the formula:
   $P = \dfrac{4\pi DF}{\theta r^3}$ where P is the pressure, D is the diameter of an expandable orifice, F is the flow rate, and r is the radius of a certain pipe. When the orifice has a diameter of 0.03 m, the flow rate is 4.5×10^{-6} m^3/s. If the orifice diameter is changed to 0.05 m, what is the new flow rate if all the other quantities remain constant?

5. Some complicated flow meter measuring device is governed by the formula:
   $P = 4\pi L\eta\dfrac{F}{r^4}$ where P is the pressure, L is the length of the main tube, F is the flow rate, r is the radius of the main pipe, and η is the viscosity of the fluid. If the pressure, length and viscosity are kept constant, but the radius is reduced by 33%, by what percent does the flow rate decrease?
   A. 20%
   B. 33%
   C. 50%
   D. 80%

6. Two plates separated by a distance of 20 cm each have a charge density of 600 C/m^2. The voltage between the plates is 500 V. Find the electric field strength between the plates.
   A. 2.5 kV/m
   B. 300 kV/m
   C. 500 kV/m
   D. 1200 V/m

7. Blood moves at a speed of 0.3 m/s through an artery with a diameter of 1×10^{-2} m. What is the volume flow rate through the artery?
   A. 3×10^{-6} m^3/s
   B. $7.5\pi \times 10^{-6}$ m^3/s
   C. $6\pi \times 10^{-5}$ m^3/s
   D. $3\pi \times 10^{-5}$ m^3/s

Answers to Questions and Problems

1. As with many problems, this one can be worked by just analyzing the units. Speed is measured in meters per second, and we are given a quantity measured in meters and a quantity measured in inverse seconds. One way of getting a quantity with the correct units is to multiply the two quantities together. This procedure happens to be the correct one. The speed of light is $(6 \times 10^{14}$ cycles/s$)(5 \times 10^{-7}$ m$) = (3 \times 10^8$ m/s$)$.

2. You should be able to immediately eliminate A (because the sum of two perpendicular vectors has to be greater than either vector) and D (because the sum has to be less than the sum of the magnitudes of the vectors). Since the vectors are perpendicular, the answer is found using the Pythagorean theorem $\sqrt{100^2 + 30^2} = \sqrt{10000 + 900} = \sqrt{10900}$. You should now be able to guess that the answer is B (104 m/s).

3. The answer has units of length so we check the units of the answers.
   A. vg has units of $[L/T][L/T^2] = [L^2/T^3]$
   B. $\dfrac{g}{v^2}$ has units of $[L/T^2]/[L/T]^2 = 1/[L]$
   C. $\dfrac{v}{g}$ has units of $[L/T]/[L/T^2] = [T]$
   D. $\dfrac{v^2}{g}$ (the reciprocal of B) has units of $[L/T]^2/[L/T^2] = [L]$
   The only answer with the correct units is D.

4. If all the other quantities remain constant then according to the formula DF must be a constant. Therefore, $D_1F_1 = D_2F_2$ Solving for F_2 we have:
   $F_2 = D_1F_1/D_2 = (0.03$ m$)(4.5 \times 10^{-6}$ m^3/s$)/(0.05$ m$) = 2.7 \times 10^{-6}$ m^3/s.

5. Since everything is constant except F and r, the formula tells us that:
   $\dfrac{F}{r^4}$ = constant. Let r be the initial value and let r_f be the final value. Since r is reduced by 33% or 1/3, the final r value must be 2/3 the initial value. Therefore: $r_f = (2/3)\, r$ And the formula $(F/r^4 = $ constant$)$ gives us:
   $$\frac{F}{r^4} = \frac{F_f}{r_f^4} \quad \text{or} \quad F_f = F\left(\frac{r_f}{r}\right)^4 = F\left(\frac{\frac{2}{3}r}{r}\right)^4 = F\left(\tfrac{2}{3}\right)^4 = F\,\frac{16}{81} \approx \frac{1}{5}F$$
   Since the final value is 1/5 the initial value, F was decreased by 4/5 or 80%. The answer is D.

6. If you know nothing about electric fields, notice that all the answers have units of V/m or kV/m. One quantity given in the problem is a voltage and another is a distance, so try the voltage divided by the distance.
   $(500$ V$)/(0.2$ m$) = 2.5 \times 10^3$ V/m $= 2.5$ kV/m. The answer is A.

7. All the answers have units of m³/s so we have to do something with the given quantities to get these units. Flow through an artery suggests that the area of the artery (in m^2) might be an important quantity, and a speed times an area has the correct units. The area of a circle is pi times one fourth the square of the diameter ($\pi D^2/4$). (Notice also that a factor of pi is in three of the answers.) The flow rate would be:

(0.3 m/s) π $(1 \times 10^{-2}$ m$)^2/4 = 7.5\pi \times 10^{-6}$ m³/s The answer is B.

2 Kinematics

2.1 Kinematics: A Description of Motion

Mechanics is the area of physics that involves the study of motion and the concepts of force and energy. **Kinematics** is the branch of mechanics that attempts to describe the motion of an object without investigating the causes of the motion. This description generally involves finding the position and velocity of the object as a function of the time. **Dynamics**, a study of the forces (pushes and pulls) that cause this motion, will be discussed later.

Often in physics we are interested in the **change** in some particular quantity. We might wonder how the temperature changes, or how the velocity changes, or how the voltage changes. Implied but not always explicitly stated is the fact that changes generally occur over time. The change of a quantity Q is always defined to be the later value (usually called the final value) minus the earlier value (or the initial value). We use the symbol "Δ" to denote a change.

$$\Delta Q = Q_{final} - Q_{initial} = Q_f - Q_i \qquad \qquad \textit{change in Q}$$

With this definition, a positive change indicates an increase in Q and a negative change tells us that Q is decreasing with time. Describing the motion of an object requires the introduction of several quantities that change with time.

The **displacement** of an object is a vector quantity defined to be the straight line distance from a starting point to an ending point. It is the change in the position of the object during a given interval of time. Since it is a vector, a full description must include the direction of the displacement. If the motion is along the x axis, the direction of the displacement is indicated by a positive or negative sign.

Example 2.1

For an initial position of $x_i = 8$ m and a final position of $x_f = 3$ m, find the displacement.

The displacement is defined as the final position minus the initial position. Therefore we have:

$$\Delta x = (x_f - x_i) = (3 \text{ m} - 8 \text{ m}) = -5 \text{ m}$$

The minus sign tells us that the displacement is in the negative x direction. (To the left if a standard coordinate system is used.) The displacement is not necessarily the actual distance covered (unless the object happened to move in a straight line without backtracking).

11

The displacement of an object from one location to another is called **translational motion**. The motion of a real object is generally a combination of translational motion and rotational motion. Rotational motion will be discussed separately.

The **average velocity** is defined to be the vector displacement divided by the time it took to get from the initial position to the final position. It is the change in position divided by the change in time. Average velocity is a vector quantity and is measured in meters per second (m/s). The average velocity when making any round trip is zero since the initial and final positions are the same (the displacement is zero).

The **instantaneous velocity** is the velocity of an object at some specific instant in time. A way to *approximate* the instantaneous velocity would be to measure the average velocity over a very short time interval, a time so short that the velocity does not change much over the time of the measurement. Ideally we would like to make the time interval infinitesimally short (although that might not be easy to do in practice). For motion along the x axis, the slope of the x versus t graph is equal to the instantaneous velocity at that time.

Speed is a term used when referring to the magnitude of a velocity vector. An object's average speed is the *actual* distance traveled divided by the time to travel that distance. The average speed is *not* the magnitude of the average velocity.

The **average acceleration** of an object is the change in the object's instantaneous velocity divided by the time over which that change took place. (A change in the instantaneous velocity is the final instantaneous velocity minus the initial instantaneous velocity.) Accelerations are measured in meters/second per second (or equivalently m/s^2). If the acceleration and velocity are in the *same* direction, the object's velocity will *increase*, but if the acceleration and velocity are in *opposite* directions, the object's velocity will *decrease*.

The **instantaneous acceleration** is the acceleration of an object at some specific instant in time. It is *approximately* the average acceleration over a very short interval of time.

An acceleration can be a change in the speed of an object (the magnitude of the velocity), a change in the object's direction of motion (the direction of the velocity vector), or a combination of both. Except for circular motion (were the direction of the velocity vector is continually changing), we will concentrate on problems where the acceleration is constant. For motion along a straight line, the slope of the v versus t graph is equal to the instantaneous acceleration at that time.

Problem Solving Strategy: When calculating a physical quantity, a valuable skill is the ability to find an approximate answer quickly. To accomplish this, round off the relevant numbers to *one* significant figure before doing the mathematical manipulations.

Example 2.2

Find the average speed of a vehicle that covers 19.3 meters in 5.17 seconds.
A. 2.35 m/s
B. 3.73 m/s
C. 4.27 m/s
D. 5.14 m/s

First, round off the distance to 20 meters and the time to 5 seconds. The average speed is approximately (20 m)/(5 s) or about 4 m/s. Since we rounded the numerator up and the denominator down, the actual answer must be somewhat *less* than 4 m/s. The answer is B.

Uniformly Accelerated Motion

Uniformly accelerated motion is motion where the acceleration is constant or can be assumed to be constant over the time of interest. Problems with constant acceleration constitute an important set of kinematics problems. The acceleration of an object is caused by the forces acting on it, and if the forces are constant, the acceleration of the object will also be constant.

The following equations describe linear motion with *constant* acceleration) along the x axis. These equations assume that the initial *position* is zero and the initial *time* is zero. Since the initial instantaneous velocity is the value at time zero, it is usually represented by the symbol v_o.

When the acceleration of an object is a non zero constant, the instantaneous velocity either increases or decreases *linearly* with the time. Therefore, the average velocity ($\bar{v}$) is just the instantaneous velocity midway between the initial instantaneous velocity (v_o) and the final instantaneous velocity (v).

$$\bar{v} = \frac{v_o + v}{2} \qquad\qquad \textit{(constant acceleration only)}$$

In terms of the average velocity, the displacement (x) at time t is given by:

$$x = \bar{v}t \qquad\qquad\qquad \textit{displacement}$$

The acceleration (a) which is constant is the change in the instantaneous velocity divided by the change in time. (The change in time is just t since we have picked zero as our initial time.) Therefore:

$$a = \frac{\Delta v}{\Delta t} = \frac{v - v_o}{t} \qquad \textit{(constant acceleration only)}$$

These three equations can be used to solve most problems with constant acceleration.

If the initial velocity is zero, the average velocity is $v/2$ (one half the final velocity) and the final velocity is the acceleration multiplied by the time. Therefore, the displacement can be written as: $x = \frac{1}{2}vt = \frac{1}{2}(at)t$. This can be written:

$$x = \frac{1}{2}at^2 \qquad \textit{(constant acceleration and initial velocity of zero)}$$

The main need for this last equation is to find the time if the distance and acceleration are known.

Falling Bodies

The downward acceleration of objects near Earth's surface is called the **acceleration due to gravity**. The magnitude of this acceleration decreases with increasing elevation, but if the altitude of a projectile does not change too much, we can assume that the acceleration due to gravity is approximately constant. Its magnitude (represented by the symbol g) is about 9.80 m/s^2.

An object whose motion is controlled only by Earth's gravity is said to be in **free fall**. It is called free *fall* even though it may be moving *up* during the first part of its journey. The object will have an acceleration of g in the downward direction. When an object is moving up, the *downward* acceleration of gravity will slow the object, and when an object is moving down, the downward acceleration of gravity will make the object go faster. The acceleration of gravity causes an object's *vertical* speed to change by 9.8 m/s each second.

If a particle is launched upward, its speed at a given elevation on its way up will be the same as its speed at that same elevation on its way down (assuming we can ignore air friction). The time it takes the object to go up a given distance is the same as the time to fall down that same distance.

If an object is *dropped* ($v_0 = 0$) from a height h, the time to fall is related to the height by the following constant acceleration equation:

$$h = \frac{1}{2}gt^2$$

14

Turning the problem around, this equation can also be used to find the time it takes for an object thrown upward to travel up to its maximum height (h). (The place where it stops before falling back down.)

Problem Solving Strategy: To obtain an approximate answer, use 10 m/s^2 for g. The velocity of an object in free fall changes (increases if falling or decreases if rising) by about 10 m/s *each second*. If you remember that this value of g is too large by about 2% the answer can be corrected if needed *after* the calculation.

Example 2.3

If a rock is dropped from rest, how far will it fall in 5 sec?
A. 5 m
B. 25 m
C. 123 m
D. 245 m

The distance it falls is given by $h = \frac{1}{2}gt^2$. Substituting for the given quantities (using $g = 10$ m/s^2) we have: $h = \frac{1}{2}(10 \text{ m/s}^2)(5 \text{ s})^2 = 125$ m. The answer is C. Our calculated answer is too large by 2% because we used 10 m/s^2 instead of 9.8 m/s^2 for g.

Graphical Representations and Summary

The instantaneous velocity at any given time is the slope of the position versus time graph at that instant. When the acceleration is constant, a plot of the position as a function of the time is a parabolic curve. The acceleration at a given time is the slope of the instantaneous velocity graph at that time. If the acceleration is constant, a plot of the instantaneous velocity as a function of the time is a straight line whose slope is the acceleration.

Our goal is to be able to solve problems where an object is moving in a straight line with a constant acceleration. The equations of motion allow us to find the position and the instantaneous velocity at any time.

As an example, we might want to describe the motion of a car as it travels down a street. As you might imagine, our task could

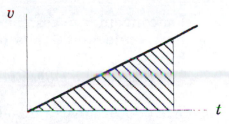

Figure 2.1 A plot of the instantaneous velocity as a function of the time for constant acceleration and initial velocity of zero. The slope of the line is the acceleration. The area under the line out to the time t_1 represents the distance traveled during that time.

be very difficult. The car may stop and start many times as it moves along the street. Generally speaking, it is impossible to accurately describe the motion of a real car with a single set of equations. Therefore, we will be happy if we can describe the car's motion over just a short portion of its trip. For example, we could describe its motion as it slows down near a school zone, or perhaps we could describe its motion as it speeds away from a stop sign. We will generally have to analyze the car's motion in segments.

Our investigation is of course motivated by the hope that we will be able to write down equations that describe the motion of cars, bullets, baseballs, rockets, and many other real-life objects. These objects might be spinning or rolling, but for the sake of simplicity, we will limit our initial discussion to objects that are not rotating. Real objects also have a physical size, but to simplify our description of where the object is located, we will pretend like they are very small, so small that we can consider them to be mathematical points. Physicists use the word **particle** to describe point-like objects.

Example 2.4

The graph is a plot of the instantaneous velocity of a car as a function of the time.

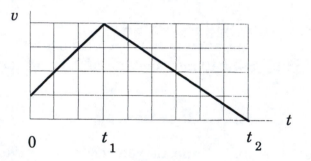

1. The ratio of the acceleration between 0 and t_1 to the magnitude of the acceleration between t_1 and t_2 is:
   A. 1:1
   B. 1:2
   C. 1:3
   D 3:2

For convenience assume each box on the graph represents 1 second on the time scale and 1 m/s on the velocity scale. The slope (acceleration) between 0 and t_1 is 1 m/s^2, and the slope between t_1 and t_2 is -(2/3) m/s^2. The ratio of the *magnitudes* is 1:2/3 = 3:2 the answer is D.

2. What does the area under the line between 0 and t_1 represent?
   A. Acceleration of the car
   B. Average velocity of the car
   C. Change in velocity of the car
   D. Distance traveled by the car

Assuming the same scale as in the last question, the acceleration is the *slope* of the line (1 m/s^2), the average velocity is the average value of the line between the two times (2.5 m/s), and the change in velocity is the

16

value at t_1 minus the value at $t=0$ (4 m/s - 1 m/s = 3 m/s). The distance traveled is the area under the line (7.5 m). The answer is D.

Two-Dimensional Motion

Two-dimensional motion is no more difficult than one-dimensional motion, but it requires a little more work, since the motion in each direction must be treated *separately*. The only two-dimensional motion we will study is **projectile motion**. Once an object is thrown or dropped, it is called a projectile if its motion is governed *only* by the pull of gravity. As soon as the object loses contact with its launching vehicle, the object is called a projectile. Projectile motion assumes Earth is flat and that air friction is small enough to be ignored. The path of a projectile is a parabola. For many objects, projectile motion is a good first approximation to reality.

The only acceleration a projectile experiences is the acceleration due to gravity, which is directed downward. When analyzing the vertical motion, the horizontal motion can be ignored since it has absolutely no influence on the vertical motion. A projectile fired off horizontally (at any speed) will hit the ground at exactly the same time as a projectile that is released from the same height at the same time. (We are assuming that the Earth is flat and that air friction can be ignored.) A projectile launched upward, will travel upward until its upward or vertical speed goes to zero.

The projectile experiences *no* acceleration in the horizontal direction, therefore, the speed in the horizontal direction does not change. The horizontal distance covered by the projectile is just the horizontal speed of the projectile multiplied by the time the projectile is in the air.

The **launch angle** is the angle the initial velocity vector makes with the horizontal. For a given launching speed, the larger the launch angle, the longer the projectile will spend in the air. The maximum **horizontal range** is realized with a launch angle of 45°. For a given launch speed, launch angles that differ from 45° by the same amount will have the same range. For example, particles launched with the same speed at 40° and 50° will have the same range since they are both 5° from the maximum range angle of 45°. However, the projectile launched at 50° will spend more time in the air since its launch angle is larger. This discussion assumes that air friction is negligible.

Example 2.5

From the top of a tall building, a bullet is fired horizontally at a speed of 350 m/s. How far does it fall in 0.01 seconds?
A. 2.4×10^{-4} m
B. 4.9×10^{-4} m

C. 5.0×10^{-4} m

D. 9.8×10^{-4} m

The horizontal speed has no effect on the vertical motion. The initial vertical speed is zero, so the distance it falls is given by $h = \frac{1}{2} g t^2$.

Using g = 10 m/s^2, we find the distance the rock falls is:

$h = \frac{1}{2}(10 \text{ m/s}^2)(0.01 \text{ s})^2 = 0.0005$ m $= 5 \times 10^{-4}$ m.

Since we used a value of g that was a little too large (by 2 %), the answer must be 2% smaller than what we calculated. The answer is B.

Circular Motion and Centripetal Acceleration

If an object is accelerating, the *magnitude* of its velocity vector may be changing or the *direction* of its velocity vector may be changing (or both may be changing). The *magnitude* of the velocity vector will change if the acceleration vector has a component that is *tangent* to the path of the particle. (It will either speed up or slow down since the acceleration vector is parallel to the velocity vector). The *direction* of the velocity vector will change if the acceleration vector has a component that is *perpendicular* to the velocity vector. An object moving in a circle must be accelerating since its direction is continually changing. The direction of the velocity vector (which is tangent to the circle), changes because the acceleration vector has a component directed toward the center of the circle (perpendicular to the velocity vector). This component of the acceleration vector is called the **centripetal acceleration**, and its magnitude depends on the speed of the object and the radius of the circular orbit. The centripetal acceleration is given by the equation:

$$a_c = \frac{v^2}{r} \qquad\qquad \textit{centripetal acceleration}$$

If the speed of an object changes as it progresses around a circle, then there must also be a component of the acceleration vector that is tangent to the circle. This component is called the tangential acceleration.

2.2 Relative Velocity

If you are paddling your canoe in a body of water (like a river) while the water is moving in some direction relative to the shore, your vector velocity *relative to the shore* is the sum of your velocity *relative to the water* and the water's velocity *relative to the shore*. We can generalize this statement as the following *vector* equation:

$$\mathbf{v}_{AC} = \mathbf{v}_{AB} + \mathbf{v}_{BC} \qquad\qquad \textit{relative velocity}$$

Where $\mathbf{v}_{AC}$ is the velocity of you (A) relative to the shore (C), $\mathbf{v}_{AB}$ is the velocity of you relative to the water (B), and $\mathbf{v}_{BC}$ is the velocity of the water relative to the shore (C). *Hint*: Solving this equation for one of the vectors will not be easy *unless* the three vectors involved form a simple triangle. Make sure you know the three simple triangles.

Example 2.6

An airplane is flying north at a speed of 40 m/s relative to the outside air, but the wind is blowing east at 30 m/s relative to the ground. How fast is the airplane flying relative to the ground?

Draw a picture to show the two velocities given. In the picture, $\mathbf{V}_{AB}$ is the velocity of the plane relative to the outside air and $\mathbf{V}_{BC}$ is the velocity of the air relative to the ground. We need to add them to find the velocity of the airplane relative to the ground which is $\mathbf{V}_{AC}$ in the picture.

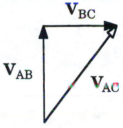

Since the two velocities we need to add are perpendicular, we can use the Pythagorean theorem to find the resultant speed. Or, we could recognize that the two vectors we are adding are the legs of a 3-4-5 triangle. The speed of the airplane relative to the ground is 50 m/s.

2.3 Problem Solving Techniques

1. Understand the problem. Take enough time when reading the problem to make sure you fully understand what you are given and what you are being asked to find. When first reading the question, do not concentrate on the actual numbers, but concentrate on what physical quantities are given (speed, mass, time, etc.). It is very difficult to decide which physical principles you will need to apply if you do not have a through understanding of the problem.

2. Some questions are so simple that they can be solved with little effort. Physics problems are not always of this type. They are rarely of the "plug into the correct equation and crank" type. You may need to give them some careful thought. Drawing a picture that includes the given quantities may be an important step in solving a problem. A picture may help you visualize the problem and get you started on the solution.

3. After calculating an answer, check to make sure your answer is reasonable. An important skill is the ability to estimate an approximate answer. If a question asks you to calculate the number of molecules in a glass of water and you get an answer of one thousand, you should realize that your answer is

absurd. If you know that Avogadro's number is about 6×10^{23}, you might guess an answer of about 10^{24} molecules before you do any calculations.

Note: On the MCAT, the symbol v may be used to represent velocity.

Questions and Problems

1. Name the three "accelerators" on an automobile.

2. A car is moving south, but it is slowing down. In which direction is the acceleration of the car?

3. A cart moves along the y axis from the $y = -10$ m point to the $y = -40$ m point. Find the displacement of the cart.

4. Suppose you leave home at 8:00 A.M. and drive to a town 180 miles away. You arrive at 11:00 A.M. and return home at 4:00 P.M. Find your average speed for the trip.

5. Suppose you leave home at 8:00 A.M. and drive to a town 180 miles away. You arrive at 11:00 A.M. and return home at 4:00 P.M. Find your average velocity for the trip.

6. If the acceleration of a block is up, in which direction is the block moving?

7. When a ball is thrown straight up in the air, the ball rises, but momentarily stops before falling back to the ground. What is the acceleration of the ball at the instant it stops at the top of its path.

8. When a ball is thrown up in the air, it momentarily stops at the top of its path. What would happen if the acceleration of the ball became zero at the top of the ball's path?

9. If an object is moving in a straight line at a constant speed, what is its acceleration?

10. Can the velocity of an object be zero at a time when its acceleration is not zero?

11. For an object moving along the x axis, if you plot the position of the object as a function of the time, how can you find the instantaneous velocity of the object at any instant in time?

12. If a graph of the instantaneous velocity as a function of the time is a straight line, what can you say about the acceleration of the object.

13. A rock is thrown horizontally off the roof of a building at the same time a bullet is fired horizontally off the roof of the building. Which will hit the ground first. Assume air friction can be neglected.

14. A car is moving along the x axis at a speed of 30 m/s in the negative x direction. If the car is located at the 5 m position in the positive x direction, write the velocity vector in unit vector notation.

15. Suppose you are given a plot of the velocity as a function of the time. The plot is a straight line that starts with a velocity of 20 m/s at $t = 0$ and ends with a velocity of 50 m/s at $t = 3$ s. What is the area under the line and what does it represent?

16. If an object is thrown vertically upward with a certain velocity, how will its velocity be different when it returns to your hand if air resistance is negligible?

17. If an object is thrown vertically upward with a certain velocity, how will its velocity be different when it returns to your hand if air resistance is not negligible?

18. A projectile is launched with a speed of 100 m/s and an angle of elevation of 30°. Its horizontal speed will be directly proportional to the:
    A. angle of elevation
    B. tangent of the angle of elevation
    C. sine of the angle of elevation
    D. cosine of the angle of elevation

19. Two identical projectiles are launched with the same initial speed but particle A is launched with a 80° launch angle and particle B is launched with a 40° launch angle. Which one covers a greater horizontal distance and which one spends more time in the air?

20. Two identical projectiles are launched with the same horizontal speed but particle A is launched with a 20° launch angle and particle B is launched horizontally. Which one covers a greater horizontal distance and which one spends more time in the air?

21. At the highest point on the path of a projectile, what quantity becomes a minimum?

22. Two identical projectiles are launched with the same initial vertical speed but particle A is launched with a horizontal speed of 20 m/s and particle B is launched with a horizontal speed of 25 m/s. Which one covers a greater horizontal distance and which one spends more time in the air?

23. An ultrasonic wave travels in the body at a speed of 1500 m/s. If a reflection is detected 8×10^{-5} sec after the wave is sent, how far below the surface is the object causing the reflection?

24. The distance between Earth and the Sun is 1.5×10^{11} meters. If one year is about 3×10^7 seconds, what is the approximate speed of the Earth about the Sun. Assume Earth's orbit is circular.

25. How long will it take a vehicle to move 2.5 m if it accelerates from rest at a rate of 20 m/s^2.

26. If a bullet is fired up at a speed of 300 m/s, how high will the bullet go?

27. A baseball is thrown horizontally off the roof of an 80 m tall building with a speed of 30 m/s. How far will it travel horizontally before it hits the ground?

28. A boat is rowed at a constant speed of 2 m/s perpendicular to the current of a river. The speed of the current is 2 m/s and the river is 200 meters wide. How far does the boat move relative to the shore before it reaches the other side?

29. A laser light pulse, which travels at 3×10^8 m/s is fired at Venus. If Venus is 7.5×10^{10} m away, how long with it take the light pulse to return?

Answers to Questions and Problems

1. The gas peddle increases the magnitude of the car's velocity vector, the brake decreases the magnitude of the car's velocity vector, and the steering wheel changes the direction of the car's velocity vector.

2. If an object is slowing down, its acceleration vector and velocity vector point in opposite directions. Therefore, the car's acceleration vector points north.

3. The displacement is the final position minus the initial position. Therefore, the displacement is: (-40 m) - (-10 m) = -40 m + 10 m = -30 m

4. The average speed is the total *distance* traveled divided by the time to travel that distance. Since the distance traveled was 360 miles and the time was 8 hours, the average speed was: (360 miles)/(8 hours) = 45 mph

5. The average velocity is the *displacement* divided by the time of the displacement. Since you are back where you started, your displacement is zero and so is your average velocity.

6. The acceleration does not tell you the direction of the motion. For example, the block could be moving down with decreasing speed. Or, it could be moving up with increasing speed, or it could be moving sideways.

7. The ball's acceleration is 9.8 m/s^2 (directed downward) for the entire time it is in the air, even when it momentarily stops. As the ball rises, the downward acceleration slows the ball by 9.8 m/s each second until it momentarily stops. The downward acceleration then causes the ball to fall with ever increasing speed. The downward speed increases by 9.8 m/s each second it falls.

8. If the acceleration went to zero when the ball momentarily stopped, the velocity would not change. (The acceleration is the change of the velocity divided by the change in time.) Therefore, the ball would remain suspended at the top of its path and never fall back down.

9. Its vector velocity is constant, so its acceleration must be zero.

10. Yes, but its velocity will not remain zero unless its acceleration is also zero. A ball thrown straight up momentarily stops at the top of its path, but its acceleration is 9.8 m/s^2 throughout its entire path.

11. The instantaneous velocity at any time is the slope of the position versus time graph at that particular instant.

12. The acceleration is the slope of the instantaneous velocity versus time graph. If the instantaneous velocity graph is a straight line, the slope (acceleration) has the same value everywhere and, therefore, must be constant.

13. At the time of launch, both objects have a *vertical* speed of zero. The horizontal speed of a projectile has no effect on its vertical motion, therefore, both objects will hit the ground at the same time.

14. The acceleration is not needed. The velocity vector is: $\mathbf{v} = -(30 \text{ m/s}) \mathbf{i}$

15. The area under the line is the average value of the line, which is the average velocity, (50 m/s + 20 m/s)/2 = 35 m/s multiplied by the time (35 m/s)(3 s) = 105 m. This product gives the distance traveled. Incidentally, the fact that the plot is a straight line tells us that the acceleration is a constant.

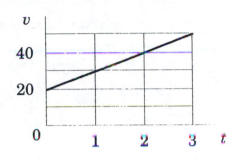

16. Its return velocity will have the *same* magnitude, but opposite direction.

17. Its return velocity will have a *smaller* magnitude and opposite direction.

18. The picture shows that the horizontal speed is: $V_{horizontal} = V \cos 30^\circ$ (It does not change during the motion.) The answer is D.

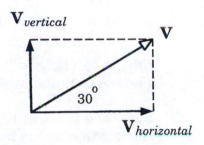

19. Particle A spends more time in the air because it has a larger initial vertical speed. However, particle B covers a greater horizontal distance since its launch angle is closer to 45°, the launch angle that gives the maximum horizontal distance.

20. Since particle A has an initial upward speed, it will spend more time in the air. Since they both have the same horizontal speed, the particle that stays in the air longest will go farther. Particle A will go farther.

21. The vertical speed goes to zero and is, therefore, a minimum. Since the horizontal speed does not change, the *total* speed is also a minimum.

22. Since their initial vertical speed are the same, they spend the same time in the air. Since particle B has a higher horizontal speed (which does not change) it will cover more ground while the two particles are in the air.

23. The time to reach the object is half the round trip time so the distance is:
(1500 m/s)(4 x 10^{-5}) = 6 x 10^{-2} m

24. (We will approximate π with 3.) The distance traveled in one year is just the circumference of Earth's circular orbit which is:
2 π (1.5 x 10^{11} m) $\approx$ 2 (3) (1.5 x 10^{11} m) = 9 x 10^{11} m
Therefore, the speed is: (9 x 10^{11} m)/(3 x 10^{7} s) = 3 x 10^{4} m/s

25. The distance in terms of the acceleration and time is gotten from the expression $x = \frac{1}{2}at^2$. Substituting in the given quantities we get:
(2.5 m) = $\frac{1}{2}$(20 m/s^2) t^2 or t^2 = (5 s^2)/20 = (1/4) s^2 and t = (1/2) s

26. We will use 10 m/s^2 for the acceleration of gravity. An acceleration of 10 m/s^2 means the velocity of the bullet will change by 10 m/s each second. (In this problem the velocity will decrease by 10 m/s each second.) Therefore, it will take (300 m/s)/(10 m/s) = 30 seconds for the bullet to stop. Its average velocity is (300 m/s + 0 m/s)/2 = 150 m/s, so the bullet will travel (150 m/s)(30 s) = 4500 meters before stopping. (Had we used 9.8 m/s^2 instead of 10 m/s^2 we would have found a distance of 4592 meters.)

27. (Let g = 10 m/s^2.) We need to know how long the ball is in the air since the horizontal distance traveled is just the horizontal speed multiplied by the time the ball is in the air. The vertical motion is governed by the acceleration of gravity. The vertical speed is zero when the ball leaves the hand of the thrower, so its vertical distance is given by: $h = \frac{1}{2}gt^2$ and the time to fall is: $t^2 = 2h/g$ = 2(80 m)/(10 m/s^2) = 16 s^2 and t = 4 s
The horizontal distance traveled in 4 s is: (30 m/s)(4 s) = 120 m

28. Since the speed of the current relative to the shore is the same as the speed of the boat relative to the water, the rowing displaces the boat 200 m across the river while the current displaces the boat 200 m down stream. The net displacement (d) is the vector sum of these two displacements, which is the hypotenuse of a right triangle with its two legs equal to 200 m. The net displacement is: 200$\sqrt{2}$ m = 283 m

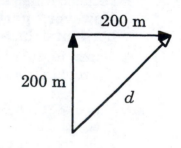

29. The total round trip distance traveled is 15 x 10^{10} m, so the time of travel is: (15 x 10^{10} m)/(3 x 10^{8} m/s) = 5 x 10^{2} s

3 Dynamics

3.1 Force and Motion

If we know the acceleration of an object and its velocity at some initial time (usually taken to be the time when our clock reads zero) we can calculate the object's displacement and instantaneous velocity at some later time. Its acceleration can be found by analyzing the forces acting on the object. Newton's second law of motion says that the vector sum of all the forces acting on an object (also known as the net force on the object) is equal to the mass of the object multiplied by the vector acceleration of the object. The net force vector and the acceleration vector point in the same direction.

Inertia is a measure of an object's tendency to remain in its present state. If it is in a state of rest, its inertia will help keep it at rest. If it is in uniform motion, its inertia will help keep it moving in a straight line at constant speed. An object's mass is a measure of its inertia.

An object's **mass** is a measure of how much resistance the object has to being accelerated. Since a bowling ball has a much larger mass than a golf ball, it is much harder to accelerate. Masses are measured in kilograms (abbreviated kg).

A **force** is a push or a pull that has the ability to change the motion of an object. The **newton** is the unit of force in the International System of Units (SI). One newton (abbreviated N) is the equivalent of one kg·m/s^2. The **net force** on an object is the vector sum of all the forces acting on the object.

The **weight** of an object is defined to be the *magnitude* of the force exerted on the object by Earth's gravity. (Sometimes we talk of the weight of an object on the Moon which is the magnitude of the force that is exerted on the object by the Moon's gravity.) The weight of an object of mass m is given by:

$$w = mg \qquad\qquad\qquad\qquad weight$$

The **tension** in a rope is the force that is exerted by the rope. When you pull on a light rope, the force of your pull is transmitted down the rope and applied to the object attached to the other end of the rope. A light rope is one whose mass is small compared to the mass of the object being pulled by the rope.

If a surface exerts a force on an object ($\mathbf{F}_S$), the *component* of the force that is perpendicular or normal to the surface is called the **normal force** ($\mathbf{F}_N$). The *component* of the force that is parallel to the surface is called the **frictional force** ($\mathbf{F}_{fr}$). (See Figure 3.1) The frictional force always opposes the object's motion or its tendency to move. If the object is not moving, the frictional force is called the force of static friction. If the object is sliding, the frictional force is called the force of kinetic friction. (It points in a direction opposite to the

27

object's velocity.) A round object rolling on a level surface (not sliding) will eventually stop. The force that causes the rolling object to stop is called the force of rolling friction.

The force of friction on a sliding object is generally proportional to the normal force on the sliding object. The **coefficient of kinetic friction** is equal to the *magnitude* of the kinetic frictional force divided by the *magnitude* of the normal force. Its value depends on the materials involved. The force of kinetic friction (f_k), the coefficient of kinetic friction (μ_k), and the normal force (F_N) are related by the following (non vector) equation:

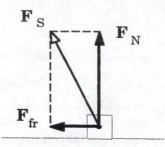

Figure 3.1 The block is sliding to the right. Only the force exerted by the surface (F_S) and its components are shown.

$$f_k = \mu_k F_N \qquad\qquad kinetic\ friction$$

The **coefficient of static friction** is equal to the maximum force of friction that can be exerted on an object before it starts to move, divided by its normal force (F_N). The force of static friction (f_s) is *less than* or equal to the product of the coefficient of static friction (μ_s) and the normal force.

$$f_s \le \mu_s F_N \qquad\qquad static\ friction$$

Air resistance (sometimes called air friction) is the force exerted on a moving object because of collisions with air molecules. Air resistance increases as the speed of the object increases. It also depends on the surface area of the object. A falling object will eventually stop accelerating as its air resistance increases. It will then continue to fall at a constant velocity called the **terminal velocity** of the object. The terminal velocity of an object generally depends on the object's weight and its surface area. If you increase your surface area (with a parachute) you will decrease your terminal velocity.

An object is said to be in **translational equilibrium** if the net force applied to it is zero. If its net force is zero, its acceleration will also be zero, but the object could still be moving with a constant velocity (moving in a straight line at a constant speed). If an object that is at rest and is also in translational equilibrium, it is said to be in static translational equilibrium. Such an object is at rest and will remain at rest as long as its net force remains zero.

Example 3.1

Air friction is approximately proportional to the square of the velocity. If the speed of a car is increased from 25 mph to 75 mph, by what factor is air friction increased?

Air friction is proportional to v^2 so it is given by an expression like: $f = bv^2$
The new velocity is 3 times the old, so the new frictional force is given by:
$b(3v)^2 = 9bv^2 = 9f$ The frictional force is increased by a factor of 9.

Newton's Laws of Motion

Newton's first law of motion (The law of inertia): *The motion of an object will not change if the vector sum of all the forces acting on the object is zero. (That is, if the net force is zero.) An object at rest will remain in a state of rest and a moving object will continue moving in a straight line with uniform speed (moving with constant velocity).*

Newton's second law of motion: *The vector sum of all the forces acting on an object (the net force on the object) is equal to the mass of the object times its vector acceleration.* ($\mathbf{F}_{net} = m\mathbf{a}$)

Newton's third law of motion (The action-reaction law): *If body A exerts a force on body B (call it an action), then body B will exert an opposing force of equal magnitude back on body A (called a reaction).*

Newton's first law is really just a special case of the second law. It says that if the force vectors acting on an object add to zero, then the acceleration must also be zero. Since the acceleration is a measure of the change in the particle's velocity, if the acceleration is zero, the velocity vector will not change. If the velocity vector does not change in direction or magnitude, the particle must either be at rest or it must be moving in a straight line with a constant speed.

Newton's second law tells us how the sum of all the force vectors acting on a particular object is related to the vector acceleration of the object. Determining the forces acting of an object is not as hard as it might first appear. Generally, in order for one object to apply a force to another object, the objects must touch.

There are two important exceptions to this "touching" rule. Objects can attract each other gravitationally without touching. If both objects are small, this attractive force can be ignored, but if one object is massive enough (like Earth, Moon, or the Sun), the force of gravity will be significant. For example, Earth's gravitational force on a baseball must be included if we are to understand the motion of the baseball.

The second exception to the touching rule involves electrical forces. Electrically charged objects can exert forces on each other without touching.

Rule for identifying forces: For an electrically neutral object on Earth, the number of forces acting on the object is equal to the number of things making contact with the object plus one. (The additional force is the gravitational attractive force of Earth, or the weight of the object.)

Example 3.2

A 10 kg box slides down a 30^0 incline. If the force of friction is 20 N, find the acceleration of the box.

There are two forces on the box (since only the incline is touching the box), but we break the force exerted by the inclined surface into two components, the normal force and the 20 N frictional force. The x coordinate must be down the incline and we resolve the force of gravity into two components. The picture is shown at the right.

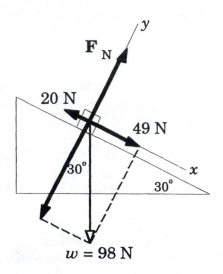

The force down the incline is 49 N and the force of friction up the incline is 20 N, therefore, the net force on the block is 29 N down the incline. Applying Newton's second law ($F_{net} = ma$) we have:
29 N = (10 kg)a
So: a = (29 N)/(10 kg) = 2.9 m/s^2

Uniform Circular Motion

Any object moving in a circle must be accelerating. The component of the acceleration that changes the direction of the velocity vector is called the centripetal acceleration. The centripetal acceleration points toward the center of the circle. If an object is moving in a circle at a *constant* speed, the *net* force on the object points toward the center of the circle and is called the **centripetal force**. The centripetal force is given by the following expression:

$$F = m\frac{v^2}{r}$$ *centripetal force*

3.2 Newton's Law of Universal Gravitation

Newton postulated that every object in the universe attracts every other object with a force whose magnitude is proportional to the product of the masses and inversely proportional to the square of the distance between them. The forces lie along the line joining the two particles. For spherical objects, the distance is measured from the center of the objects. The magnitude of the force of attraction between masses m_1 and m_2 is given by:

$$F = G\frac{m_1 m_2}{r^2}$$ *Newton's law of gravity*

Where r is the distance between the masses and G is a constant known as the universal gravitational constant.

Near the surface of Earth, the weight of a particle is the force of attraction between Earth and the particle. If m is the mass of the particle and w is its weight, we have:

$$w = mg = G\frac{mm_e}{r_e^2}$$ *weight*

And we see that:

$$g = G\frac{m_e}{r_e^2}$$ *acceleration of gravity*

Where m_e is the mass of Earth and r_e is its radius.

The Moon or an artificial satellite is able to orbit Earth because the force of gravity supplies the centripetal force that is necessary to change the direction of the velocity vector of the satellite. If the orbit is circular, or nearly so, the centripetal force will be equal to the force of gravity.

$$m\frac{v^2}{r} = G\frac{mm_e}{r^2}$$

This equation can be solved to find the velocity of the satellite in terms of its distance from Earth's center (r).

$$v = \sqrt{G\frac{m_e}{r}}$$ *orbital velocity*

This velocity does not depend on the mass of the satellite. Although many orbits are nearly circular, the general shape of an orbit is elliptical. Any object orbiting a larger body will orbit on an elliptical shaped orbit with the larger body located at one of the focal points of the ellipse. A circle is a special case of an ellipse with the two focal points located at the center of the circle. The speed of the orbiting body varies with the distance, moving *faster* as the two objects get *closer* together.

Example 3.3

If M is the mass of Earth, m is the mass of a satellite, r is the radius of the satellite's circular orbit, and G (= 6.6 x 10^{-11} N·m^2/kg^2) is the gravitational constant, the satellite's period (time to orbit Earth) is given by the following expression:

A. $2\pi \dfrac{r^3}{GM}$

B. $2\pi \sqrt{\dfrac{r^3}{GM}}$

C. $2\pi \dfrac{r^3}{Gm}$

D. $2\pi \sqrt{\dfrac{r^3}{Gm}}$

The answer must have units of time [T]. Check the units of the answers. Remember that a newton (N) has units of [ML/T^2] so G has units of: [ML/T^2][L^2]/[M^2] = [L^3/MT^2].
A and C have units of: [L^3]/[L^3/MT^2][M] = [T^2]
A comparison of A and B should be enough to recognize that B and D have units of time. (The units of B and D are the square root of the units of A and D.) The period of the satellite obviously depends on the mass of Earth (it does not depend on its own mass), therefore, the answer is B.

3.3 Equilibrium

Newton's second law says that the net force acting on an object is proportional to the object's acceleration. Often the net force acting on an object is found to be zero, and in these instances the acceleration must also be zero. A zero acceleration means that the object's velocity is a constant. In other words, the object must be at rest or must be moving at a constant speed in a straight line. An object at rest is said to be in **static equilibrium**, and an object moving at a constant speed in a straight line is said to be in **dynamic equilibrium**.

Questions and Problems

1. If an object is not moving, does it have any forces acting on it? Explain.

2. If you were traveling toward a distant star and you ran out of fuel, would your space ship slow down and stop? Explain.

3. An action hero holding a powerful gun in each hand, blasts a villain. When the projectile hits the villain, it knocks him back into the next county (almost). How does this scene violate Newton's third law of motion?

4. If an object is falling straight down at a constant speed, what is its acceleration?

5. If a 5 kg block is sliding along a level surface at a constant speed of 20 m/s and the coefficient of kinetic friction is 0.25, find the net force on the block.

6. An instant after you let go of a 10 kg brick, what will be the net force on the brick? Ignore air friction.

7. If you kick a foam ball, you will hardly feel it. However, if you kick a bowling ball you may break your toe. What would happen if you performed this experiment out in space where both are weightless?

8. A 1000 newton person jumps out of an airplane. The only force on the person is the downward force of 1000 newtons exerted on the person by the Earth. (Neglect air resistance.) Explain where the reaction force is.

9. A net force of 100 newtons to the left is being applied to a brick. In which direction is the brick moving? Be careful!

10. Seven or eight seconds after jumping out of an airplane, a person reaches his terminal velocity. If you forgot your parachute, what could you do to reduce your terminal velocity?

11. The force of gravity is given by: $F = G\dfrac{m_1 m_2}{r^2}$ The radius of Earth is 6×10^6 m. The gravitational force on a particle that is 6×10^6 m above the Earth's surface is what percentage of the gravitational force when that particle is on Earth's surface?

12. In an accelerating elevator, your "apparent weight" is the force with which you push down on the floor of the elevator. How does this force compare with the force of the floor pushing up on you?

13. A box of mass m is sliding along a level surface. Only the surface is touching the box. Let the positive x axis lie in the direction the box is sliding. The coefficient of kinetic friction between the box and the surface is μ_k. The acceleration of the box is given by:

A. $(-g + \mu_k)\,\mathbf{i}$

B. $(-g/\mu_k)\,\mathbf{i}$

C. $g\mu_k\,\mathbf{i}$

D. $-g\mu_k\,\mathbf{i}$

14. A 12 kg block on a level surface is moved by applying a horizontal force of 150 newtons. Find the force of kinetic friction if the block is accelerating at a rate of 5 m/s^2.

15. If M is the mass of Earth, m is the mass of a box, r is the radius of Earth, and G (= 6.6 x 10^{-11} N·m^2/kg^2) is the gravitational constant, the acceleration of the box due to gravity near the surface of the Earth is given by the following expression:

A. $G\dfrac{M}{r^2}$

B. $G\dfrac{M}{r}$

C. $G\dfrac{Mm}{r^2}$

D. $G\dfrac{Mm}{r}$

16. On a level surface, an 10 kg ball rolling with a speed of 20 m/s eventually stops after rolling 40 meters. Find the force of rolling friction on the ball.

17. If a net force of 3 N is applied to a 0.06 kg particle initially at rest, how long will it take for the particle to move 1 m?

18. A 2000 kg airplane is flying at a fixed elevation. Find the acceleration of the plane if its engines provide a forward thrust of 7000 N and air friction creates an opposing force of 6000 N.

19. A 100 kg man is standing on a bathroom scale in an elevator. What will the scale read if the elevator is accelerating upward at a rate of 2 m/s^2? The scale reading is sometimes called the "apparent weight" of the person.

20. If M is the mass of Earth, m is the mass of a satellite, r is the radius of the satellite's circular orbit, and G (= 6.6×10^{-11} N·m^2/kg^2) is the gravitational constant, the speed of the satellite in its orbit is given by the following expression:

A. $\dfrac{GM}{r}$

B. $\sqrt{\dfrac{GM}{r}}$

C. $\dfrac{Gm}{r}$

D. $\sqrt{\dfrac{Gm}{r}}$

21. Which of the following graphs best illustrates how the acceleration due to gravity changes with distance above the Earth's surface increases?

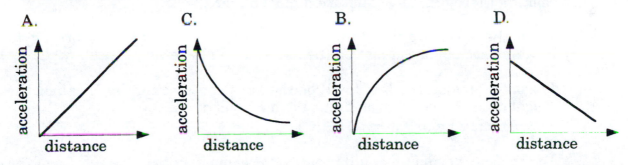

A.   C.   B.   D.

Answers to Questions and Problems

1. It could have many forces acting on it. However, the vector sum of these forces (the net force on the object) must be zero.

2. No, Newton's first law says a moving object will continue moving in a straight line at a constant speed unless some outside force causes a change in its motion.

3. The projectiles must carry quite a wallop. In order to project them at the villain, the guns must have applied a huge force to each projectile. Newton's third law says that our action hero must have received an equal force in the opposite direction. Why was our hero not thrown into the opposite county by the recoil of the guns?

4. An object moving in a straight line at a constant speed has a constant velocity, therefore, its acceleration must be zero.

5. Since the speed is constant, the acceleration is zero. According to Newton's second law of motion the net force will be zero.

6. The only force on the brick will be the force of attraction by Earth or the weight of the brick. Therefore the net force will be:
$w = mg = (10 \text{ kg})(9.8 \text{ m/s}^2) = 98 \text{ N}$

7. The space experiment would produce the same results. The inertia of the bowling ball will the same in outer space, and it will be just as difficult to get it moving. It must obey Newton's first law everywhere in the Universe.

8. The reaction force is the upward pull the person exerts on the Earth. The person will attract the Earth with a force of 1000 N. This upward force on the Earth, however, will produce a negligible acceleration of Earth.

9. You cannot tell. The direction of the net force and acceleration are the same (to the left), but they do not give the direction of the velocity. An object can be moving in one direction, but have an acceleration in another direction. For example, the brick could be moving to the right, but be slowing down (since the acceleration is pointing left).

10. Make your surface area a large as possible by spreading out your arms and legs. Opening up your shirt might help too, although probably not enough unless you are lucky enough to land in a very soft area.

11. The value of r (the distance from the Earth's center) for the distant position is 2 times the value of r on the Earth's surface. The force is, therefore, 1/4 the force on the Earth's surface or 25% of the value on the Earth's surface.

12. According to Newton's third law they are equal in magnitude.

13. The magnitude of the force of kinetic friction is μ_k times the normal force. On a level surface with no other forces acting, the normal force is numerically equal to mg. Therefore, the magnitude of the frictional force is $\mu_k mg$, and its direction is opposite the direction the box is moving (the negative x direction or $-\mathbf{i}$ direction). Since this force is the only horizontal force on the box, the magnitude of the acceleration is $F/m = \mu_k g$. The answer is D.

14. Since the 12 kg block is accelerating at a rate of 5 m/s^2, the net force on the block must be: $F_{net} = (12 \text{ kg})(5 \text{ m/s}^2) = 60$ N. There are only two horizontal forces, the 150 N force pointing in the direction of the acceleration and the force of friction (F_{fr}) which is pointing in the opposite direction. Applying Newton's second law ($F_{net} = ma$) we have:

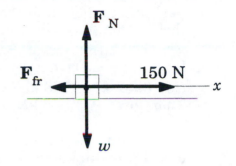

$(150 \text{ N}) - F_{fr} = (12 \text{ kg})(5 \text{ m/s}^2) = 60$ N
And solving for the force of friction:
$F_{fr} = 150 \text{ N} - 60 \text{ N} = 90$ N

15. You are not expected to remember the answer to questions like this one. All the answers obviously have different units so the right answer is the one with the correct units. Checking units we have:
A. $(\text{N·m}^2/\text{kg}^2)(\text{kg})/(\text{m}^2) = \text{N/kg} = \text{m/s}^2$
B. $(\text{N·m}^2/\text{kg}^2)(\text{kg})/(\text{m}) = \text{N·m/kg} = \text{m}^2/\text{s}^2$
C. $(\text{N·m}^2/\text{kg}^2)(\text{kg}^2)/(\text{m}^2) = \text{N} = \text{kg·m/s}^2$
D. $(\text{N·m}^2/\text{kg}^2)(\text{kg}^2)/(\text{m}) = \text{N/m} = \text{kg/s}^2$
A is the only answer with the right units.

16. The ball's average speed is $(20 \text{ m/s})/2 = 10$ m/s. Since the distance traveled is 40 m, the time to stop is 4 s (the distance divided by the average speed). The acceleration is the change in velocity divided by the change in time, so the magnitude of the acceleration is $(20 \text{ m/s})/(4 \text{ s}) = 5 \text{ m/s}^2$. The fictional force is the only horizontal force, so it is the net force. Applying Newton's second law ($F_{net} = ma$) we have: $F_{fr} = (10 \text{ kg})(5 \text{ m/s}^2) = 50$ N

17. The acceleration of the particle is $(3 \text{ N})/(0.06 \text{ kg}) = 50 \text{ m/s}^2$. The distance, time, and acceleration are related by: $x = \frac{1}{2}at^2$

Therefore: $t = \sqrt{2x/a} = \sqrt{(2 \text{ m})/(50 \text{ m/s}^2)} = \sqrt{1/25} \text{ s} = (1/5) \text{ s} = 0.2$ s

18. The forces on the plane are shown in the picture. The lift on the wings is balanced by the weight of the plane. The net horizontal force is:

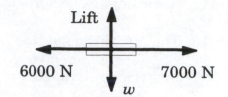

7000N - 6000 N = 1000 N

Using $F_{net} = ma$ we obtain:

$a = (1000 \text{ N})/(2000 \text{ kg}) = 0.5 \text{ m/s}^2$

19. The force of the man on the scale is equal (but pointing in the opposite direction) to the force of the scale on the man. There are two forces on the man (see picture), the scale pushing up (F_N), and the Earth pulling down. The force of the Earth pulling down on the man is his weight which is: $w = (100 \text{ kg})(9.8 \text{ m/s}^2) = 980 \text{ N}$. Applying Newton's second law ($F_{net} = ma$) we have:

$F_N - (980 \text{ N}) = (100 \text{ kg})(2 \text{ m/s}^2) = 200 \text{ N}$

Therefore: $F_N = 980 \text{ N} + 200 \text{ N} = 1180 \text{ N}$

20. From dimensional analysis, we can discover that the answer has to be either B or D. The speed of the satellite obviously depends on the mass of Earth (it does not depend on its own mass), therefore, the answer is B.

21. The acceleration due to gravity is inversely proportional to the square of the distance from the center of the Earth. Therefore, from the Earth's surface, C is the best illustration.

4 Energy, Momentum, and Torque

4.1 Work and Energy

Determining how a physical system evolves in time can be accomplished by analyzing the forces acting on the system, but it is sometimes more convenient or easier to analyze the energy of the system. For one thing, energy is a scalar quantity, but forces are vector quantities. It is much easier to manipulate scalars than vectors.

Work is something that is done to an object by a force. A force applied to an object does work *if* the object moves, even if the motion is not entirely due to that particular force. The work done by the force **F** as the object is displaced a distance **d** is given by:

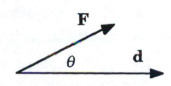

$$W = F\, d\, \cos \theta \qquad \text{\textit{work done by force} } \mathbf{F}$$

Where F and d are the *magnitudes* of the force vector and displacement vector respectively, and θ is the angle between the two vectors. The unit of work and energy in the International system of units is the joule (abbreviated J). One joule is equal to one newton-meter.

Example 4.1

How much work is done by a constant horizontal force of 40 N that pushes a 30 kg box 10 meters along a horizontal surface in 12 seconds?

The work is the force times the parallel distance, the other information is not needed. Therefore the work is: $W = F\, d \cos 0^\circ =$ (40 N)(10 m) = 400 J

Generally speaking, energy is something a body has that enables it to do work. We define several forms of energy. The **kinetic energy** is the energy a body has by virtue of its motion. The kinetic energy of an object of mass m and velocity v is given by:

$$K = \tfrac{1}{2} m v^2 \qquad\qquad \text{\textit{kinetic energy}}$$

Potential Energy

Potential energy is the energy a body has by virtue of its *position*. Two important forms of potential energy are gravitational potential energy and the potential energy stored in a spring. Potential energies are always given relative to some reference position where the potential energy is zero. Forces with which we can associate a potential energy (like the force of gravity and the force exerted by a spring) are called **conservative forces**. Other forces (like the force of friction or a force exerted by a person) are called **nonconservative forces**.

Gravitational potential energy is always measured relative to some zero energy point (often taken to be the surface of Earth). The gravitational potential energy of an object of mass m is given by:

$$U = mgh \qquad\qquad\qquad \textit{gravitational potential energy}$$

Where h is the *elevation* of the object relative to the zero position and g is the acceleration due to gravity.

The magnitude of the force exerted by a spring is proportional to the distance the spring has been stretched or compressed from its relaxed position. This relationship is known as Hook's law and is written as:

$$F = -kx \qquad\qquad\qquad\qquad\qquad \textit{Hook's law}$$

Where k is the spring constant (measured in N/m) and x is the displacement of the spring from its equilibrium position (unstretched position). The spring constant tells how much force would be required to stretch or compress the spring a distance of one meter. The minus sign is included because the direction of the force is *opposite* to the direction of the displacement of the spring.

Elastic potential energy or the energy stored in a spring is defined to be zero when the spring is relaxed. The potential energy stored in a stretched or compressed spring is given by:

$$U = \tfrac{1}{2} kx^2 \qquad\qquad\qquad \textit{potential energy of a spring}$$

Where x is the distance the spring has been stretched or compressed from its equilibrium position.

Example 4.2

If one end of a spring is held firmly and 100 g is attached to its free end, the spring stretches 0.15 m. What is the value of the spring constant?

The force of the spring is described by Hook's law: $F = kx$. The force is equal to the *weight* of the 100 g *mass*:
$F = W = mg = (0.100 \text{ kg})(9.8 \text{ m/s}) = 0.98$ N
Therefore: $k = F/m = (0.98 \text{ N})/(0.15 \text{ m}) = 6.5$ N/m

The Conservation of Energy

One of the most important principles in physics is the conservation of energy. It says that the amount of energy in an isolated or **closed system** remains constant in time. The energy may change from one *form* to another, but the total amount of energy does not change. This principle allows us to understand certain systems by analyzing the energy.

Friction exerted on a system by outside sources will introduce energy, but if the friction is small enough to be ignored, the conservation of energy principle can be applied to a system that is not completely closed.

The sum of the potential energy and kinetic energy of a system is called the total **mechanical energy** of the system. If there are no nonconservative forces present (like friction) or if the nonconservative forces can be neglected, the total mechanical energy of the system will be conserved.

A projectile and the Earth form an isolated system, and we can understand the motion of the projectile by applying the principle of energy conservation to this system. As the projectile changes elevation, its kinetic and gravitational potential energies change, but its total mechanical energy remains constant (if we can ignore air friction). Therefore, the total mechanical energy at one time (the initial time) will equal the total mechanical energy at some later time (the final time). We can write the conservation of mechanical energy as an equation:

$$K_i + U_i = K_f + U_f \qquad \textit{conservation of mechanical energy}$$

When trying to find the speed of a particle that is dropped or thrown from a given height, it is much easier to use the conservation of mechanical energy than the kinematics equations. A particle dropped from rest (an initial kinetic energy of zero) from a height h will have a speed v just before it hits the ground (where the potential energy is zero). The conservation of mechanical energy equation gives:

$$0 + mgh = \tfrac{1}{2} mv^2 + 0$$

or

$$v = \sqrt{2gh}$$

Example 4.3

A 2 kg projectile is launched upward at an angle of 60° above the horizontal with a kinetic energy of 900 J. Assuming air friction can be ignored, determine how much kinetic energy the projectile has at the highest point in its path.

Since the launch angle is 60°, the horizontal speed (which does not change) is one half of the launch speed. ($v_{horizontal} = v \cos60° = v/2$) Since the kinetic energy is proportional to the square of the speed, the kinetic energy at the highest point is:

$(1/2)^2 = 1/4$ of the initial kinetic energy, or (900 J)/4 = 225 J

A body's **internal energy** (sometimes called thermal energy) is the total *microscopic* kinetic energy of the particles (atoms or molecules) that make up the object. The temperature of a *gas* is a measure of the average translational kinetic energy of the particles that make up the gas.

Other forms of energy are electrical, nuclear, and the chemical energy stored in food and fuels. Energy that flows from one place to another because of a difference in temperature is called **heat**.

4.2 Power

Power is a measure of the *rate* at which work is done or the rate at which energy is transformed from one form to another. The average power delivered by a given force is the work done by that force divided by the time during which the work was done. In equation form this is written as:

$$P = \frac{W}{t} \qquad\qquad\qquad power$$

Power is measured in watts (abbreviated W). A rate of one watt is equivalent to a rate of one joule per second.

Example 4.4

A 0.1 kg particle is accelerated from rest to a speed of 10 m/s in 2 seconds. Find the average power delivered to the particle.

To find the power, we need the work, but the work requires the net force and the distance traveled. To find the net force we need the acceleration.

Acceleration = (change in velocity)/(time) = (10 m/s)/(2 s) = 5 m/s^2
Net force = (mass)(acceleration) = (0.1 kg)(5 m/s^2) = 0.5 N
Distance = (average speed)(time) = (5 m/s)(2 s) = 10 m
(The particle started at rest, so its average speed is half its final speed.)
Work = (force)(distance) = (0.5 N)(10 m) = 5 J
Power = (work)/(time) = (5 J)/(2 s) = 2.5 W

4.3 Momentum

One of the most important discoveries in physics is that certain physical quantities are conserved in *closed* systems. Knowing which quantities do not change in a closed system is an important tool when analyzing these systems. One of these conserved quantities is energy and another is linear momentum.

The **linear momentum** of a particle is defined to be the mass of the particle multiplied by the vector velocity of the particle. The linear momentum of a particle is a vector quantity and it is defined by:

$$\mathbf{p} = m\mathbf{v} \qquad\qquad\qquad \textit{linear momentum}$$

The **conservation of linear momentum principle** states that the total vector momentum of a *closed* system does not change with time. This principle is especially useful when analyzing systems involving collisions, explosions, or other things that happen quickly. If something happens quickly (even in a system that is not closed), there is not enough time for external forces to change the momentum of the system. The principle of the conservation of momentum says that the *vector* sum of the momenta of all the particles in a system just before a collision or explosion is equal to the vector sum of the momenta of all the particles just after the event. If we use capital letters to represent the vector sum of a system, this statement can be written as:

$$\mathbf{P}_{\text{initial}} = \mathbf{P}_{\text{final}} \qquad\qquad\qquad \textit{conservation of momentum}$$

Example 4.5

A 100 kg man is in a 150 kg boat floating at rest in a lake. If the man jumps off the boat with a horizontal speed of 6 m/s, find the speed of the boat immediately after the man jumps off the boat.

Momentum must be conserved and the initial momentum is zero. Therefore, the momentum of the man in one direction must be numerically equal to the momentum of the boat in the opposite direction. This will assure that the two final momentum *vectors* (the man's momentum and the boat's momentum) will add to be zero. If we let V be the speed of the boat just after the man jumps off we have: (100 kg)(6 m/s) = (150 kg) V
Therefore: $V = 4$ m/s

Impulse

Newton's second law can be written in terms of a change in the momentum of the particle:

$$\mathbf{F} = \frac{\Delta \mathbf{p}}{\Delta t}$$

If we multiply both sides by Δt we have an equation for the change in the momentum of a particle. An **impulse** is a force applied to an object over a short period of time. This force of short duration will change the momentum of the particle to which it is applied. A baseball bat gives an impulse to a baseball, and by Newton's third law, the baseball will give an equal and opposite impulse to the baseball bat. During a real impulse, the force changes with the time, but we can represent an impulse as an average force $\overline{F}$ over a short time interval Δt. The impulse will produce a change in the momentum of the particle Δp.

$$I = \overline{F} \ \Delta t = \Delta p \qquad\qquad\qquad impulse$$

Equal impulses can be generated by either applying a *large* force for a *short* time or by applying a *smaller* force for a *longer* time. If two cars have the same mass and are traveling at the same speed when they run into a wall, they both receive the same impulse. However, the car that collapses the most will receive its impulse as a *smaller* force over a *longer* time. The advantage of a car with a collapsible front end should be obvious.

Collisions

Momentum is a useful quantity to use when studying collisions because the total momentum is conserved in a collision and it can usually be calculated easily. Energy is also conserved, but the kinetic energy present before the collision can be changed into forms that are hard to keep track of. For example, some of the kinetic energy can be converted to heat and sound energy. Bouncy collisions, where very little kinetic energy is lost are called elastic collisions. In **perfectly elastic collisions**, no kinetic energy is lost. Perfectly elastic collisions only occur on a microscopic scale between elementary particles, but some collisions between very hard objects can be approximated as perfectly elastic.

In some collisions, the objects stick together. Such collisions are called **perfectly inelastic collisions**. Kinetic energy is *always lost* in a perfectly inelastic collision, but the total momentum is the same before and after the collision, and it can be used to analyze these collisions.

Without going through the math, we will state the results of a few common types of *head on* collisions. The results of most of these examples should be familiar to you from experience.

If a ball moving to the right has a head on collision with an *identical* ball at rest, the first ball will stop and the other ball will move off to the right with the same velocity the first ball had before the collision. This is a familiar collision to anyone who has ever played pool.

If a small mass ball moving to the right has a head on collision with a large mass ball at rest, the small ball will rebound in the opposite direction and the large ball will move off to the right with a smaller speed than the small ball had before the collision.

If a large mass ball moving to the right has a head on collision with a small mass ball at rest, both balls will move to the right after the collision, but the small ball will have a larger speed. The speed of the large ball will be less than its original speed.

Example 4.6

A 2 kg ball at rest is accelerated at a rate of 5 m/s^2 for 3 seconds. It then collides with a 3 kg ball that is at rest. If the two balls stick together, find the speed of the balls after the collision.

Momentum is conserved so we need to know the momentum just before the collision. The speed of the 2 kg ball just before the collision is:
(5 m/s^2)(3 s) = 15 m/s and its momentum is: (2 kg)(15 m/s) = 30 kg·m/s
The momentum after the collision is equal to the combined mass of 5 kg (call it M) times the velocity (call it V) of the combined mass.
Therefore: MV = 30 kg·m/s
Solving for the velocity we have: V = (30 kg·m/s)/(5 kg) = 6 m/s

4.4 Center of Mass

The **center of mass** of an object is that point where the mass can be considered to be concentrated. For a symmetrical object, the location of the center of mass can generally be found by inspection.

The motion of a real object with a physical size may be thought of as a combination of translational motion and rotational motion. The translational motion is the motion of the object's center of mass from one location to another, and the rotational motion is the rotation of the object about its center of mass.

4.5 Torque

Some forces exert a twist on an object, causing that object to spin. The amount of twist exerted by such a force is called a torque. A **torque** produced by a given force is defined to be the product of the force and the perpendicular distance from the object's axis of rotation to the line along which the force acts. (See figure to the right.) We write this definition in equation form as:

$$\tau = r_\perp F \qquad\qquad torque$$

Where F is the force and $r_\perp$ is called the **lever arm**, the length of the line segment that is perpendicular to both the rotation axis and to the line of action of the force. There is a direction associated with a torque and it is a vector quantity. For an object confined to rotate about a fixed axis, the torque can be described as either clockwise or counterclockwise.

The units of torque are m·N, which are also the units of work and energy. When discussing the *scalar* quantities of work and energy, we defined one joule to be one N·m. However, since torque is a *vector*, we do not use joules to measure units of torque. We simply leave the units of torque as m·N.

If an object is not moving in any way and is not going to move, it is said to be in **static equilibrium**. In order for an object at rest to be in static equilibrium, the net force acting on the object must be zero (the object must be in static translational equilibrium) and the net torque on the object must also be zero (the object must be in **rotational equilibrium**). The net torque on an object is zero when the clockwise torques about *any* point are equal to the counterclockwise torques about that same point.

When torques are applied to real objects, the weight of the object may apply a torque. Since the weight of an object acts as though it is concentrated at the center of mass of the object, the weight cannot apply a torque about the center of mass, however, the weight will apply a torque about all other points. For example, if a stick is held at one end, its free end will tend to rotate downward because the weight of the stick applies a torque about the end point at which the stick is held.

Example 4.7

A 50 kg uniform beam is 10 m long and is being suspended by a cable 3 m from the left end. A 20 kg mass is hanging 2 m from the right end. In order for the beam to be in equilibrium, another mass is placed 1 m from the left end. Find the mass that is placed 1 m from the left end.

About any point, the clockwise torques must equal the counterclockwise torques. Since we do not know the tension in the cable and we are not asked about it, the point where the cable is attached is the point about which we should take the torques. (The torque due to the cable will be zero about this point.) There are three torques about this point, two clockwise torques and one counterclockwise torque due to the unknown mass. The picture to the right will help us visualize the situation.

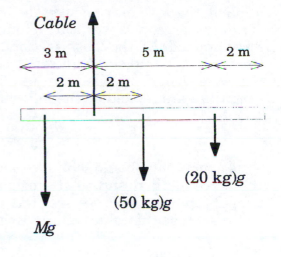

The (20 kg)g *weight* has a 5 m lever arm, the *weight* of the beam has a 2 m lever arm, and the *weight* of the unknown mass (Mg) has a 2 m lever arm. Therefore, equating the clockwise torques to the counterclockwise torques we have:
(50 kg)g(2 m) + (20 kg)g(5 m) = Mg(2 m)
Dividing out the g in each term and solving for M we have: $M = 100$ kg

4.6 Pulleys, Levers, and Inclined Planes

Pulleys, levers, and inclined planes can be used to gain a mechanical advantage. Even if you are not strong enough to lift a car, you can use a set of pulleys, a lever, or and inclined plane to lift the car. If we assume that friction is negligible, the work done in lifting the car must be done by you, but since work is a force applied over a distance, perhaps you could create a device that would allow you to do the work by applying a *smaller force* over a *longer distance*.

Example 4.8

You wish to lift a 3000 pound car up 4 feet. If an incline rises 1 meter over a distance of 50 meters, how hard would you have to push on the car to move it up the incline until it is 3 feet above the ground?

The work needed to lift the car 4 feet is:
$W = Fd = $ (3000 lb)(4 ft) = 12,000 ft·lb You must do this work by pushing the car 200 ft (50 ft for every 1 ft rise since the ratio of run to rise is 50:1). The force you need to exert (assuming no friction) is:
$F = W/d = $ (12,000 ft·lb)/(200 ft) = 60 lb

4.7 Stress and Strain

When a force is applied to a real object, but the object does not *appear* to move, the force can still have an effect on the object. For example, if a weight is hung from a wire, it may appear that nothing happened, but careful measurements will reveal that the wire stretched by a small amount. If enough force is applied, the wire may be stretched permanently or may even snap. If the force is not too large, however, the amount the wire stretches (ΔL) will depend on the force applied (F), the cross sectional area of the wire (A), the original length of the wire (L_o), and some quantity that depends on the material used to make the wire (E). E is called the **elastic modulus** or **Young's modulus**. The quantities are related by the following expression:

$$\Delta L = FL_o / EA$$

This expression should look reasonable. The stretch is directly proportional to the force applied and to the original length of the wire, but it is inversely proportional to the cross sectional area of the wire (how fat the wire is). We can rearrange the equation and write it as:

$$\frac{F}{A} = E \frac{\Delta L}{L_o}$$

The force per unit area in is equation (F/A) is called the **stress** (tensile stress in this situation), and the stretch per unit length ($\Delta L / L_o$) is called the **strain**. We say the stress is directly proportional to the strain, and the constant of proportionality is the elastic modulus or Young's modulus.

If we place a fluid in a cylinder with a piston on top and apply a downward force on the piston, the fluid will get squashed and its volume will decrease by a small amount (ΔV). We can write an equation similar to the one above for this situation:

$$\Delta V = - FV_o / BA$$

or

$$\frac{F}{A} = -B \frac{\Delta V}{V_o}$$

The stress (F/A) is the *increase* in pressure on the fluid that causes the squashing (usually written as ΔP), and the constant of proportionality B is called the **bulk modulus** of the fluid. The negative sign is necessary because an *increase* in the pressure causes a *decrease* in the original volume (V_o). This expression is generally written as:

$$\Delta P = -B \frac{\Delta V}{V_o}$$

Example 4.9

Two wires made of the same material are the same length, but wire A has twice the diameter of wire B. If equal masses are hung from the wires, how does the amount that wire A stretches compare with the amount wire B stretches?

The stretch (change in length) is directly proportional to the length and force (both of which are the same for the wires), and inversely proportional to the cross sectional area. Wire A has twice the diameter of wire B so it has 4 times the cross sectional area of wire B. Therefore, the thinner wire (wire B) will stretch 4 times more than the thicker one (wire A).

Questions and Problems

1. If the velocity of a car is doubled, by what factor is the kinetic energy changed?
   A. 1/2
   B. 2
   C. 1/4
   D. 4

2. When a projectile returns to the elevation from which it was launched, how will its kinetic energy compare with the value it had when it was launched? Assume there is air friction present.

3. A golf ball is hit off the tee with a nine iron. Its path is very high. Which of the following quantities remains constant while the ball is in the air? Assume no air resistance.
   A. kinetic energy of the ball
   B. potential energy of the ball
   C. momentum of the ball
   D. horizontal speed of the ball

4. If a rock is dropped from a height of 2 meters, about how fast will it be moving just before it hits the ground?
   A. 2 m/s
   B. 4 m/s
   C. 6 m/s
   D. 8 m/s

5. A block is sent down a 20° incline with an initial speed of 15 m/s. The work done by friction as the block slides down the 5 m long incline is W. If the speed of the block is doubled to 30 m/s, the work done by friction as the block slides down the 5 m long incline will be:
   A. 4W
   B. 2W
   C. W
   D. W/2

6. As a real projectile travels up, what energy transformations take place?

7. A person applies a horizontal 40 N force to a 20 kg block, causing it to slide 10 m along a level surface in 5 seconds. Find the power supplied by the person to push the block

8. A 2 kg projectile is launched straight up with an initial kinetic energy of 900 J. Assuming air friction can be ignored, determine how high will it rise?

9. A 50 kg box is elevated 2 meters by using a double pulley block and tackle. This device consists of two pulleys and a rope. The box is raised by pulling the rope with a force F at an angle of $60°$ with the vertical. Assuming friction forces can be ignored and the pulleys are massless, the approximate work done by the force F is:
   A. 2000 J
   B. 1000 J
   C. 870 J
   D. 500 J

10. A ball dropped from a height of 2 meters bounces and takes 1/2 second to reach its maximum height on the rebound. What percentage of its original potential energy was lost in the collision with the floor?

11. A 4 kg projectile is launched straight up with an initial speed of 30 m/s. Find the maximum height reached by the projectile.

12. A spring with 0.3 kg hanging on its end is 0.400 m long. If the 0.3 kg mass is replaced by a 0.5 kg mass, the spring's length is 0.405 m. Find the spring constant.

13. A certain thermonuclear explosion released 150 megawatt·years of energy. If the energy could have been spread out evenly over 30 years, how much power could have been produced?
   A. 5 MW/year
   B. 5 kW/year
   C. 5 W/year
   D. 5 MW

14. When you jump up and dunk a basketball, you must come down. You will hit the floor with a certain momentum, which must change to zero. Your change in momentum is accomplished by an impulse given to you by the floor. Of course you bend your knees to "break" the fall, but it also helps to have shoes that have a lot of cushioning. Exactly what does the cushioning do? Explain in terms of impulses.

15. A particle with a very small mass collides with a very massive particle that is initially at rest. Approximately how much kinetic energy is lost in the collision if the particles stick together?

16. If you have ever see a collision in a car race, you many have noticed many large pieces flying off the race car. What is the advantage to the driver of the car if several large pieces fly off his car? (The flying pieces may not be an advantage to other drivers who might get hit by the pieces.)

17. Some safety features on a car include collapsible bumpers, collapsible steering columns, and engines that get pushed downward below the passengers. What is the advantage of all these collapsing parts?

18. A 1000 kg car moving north at 20 m/s collides with a 1500 kg truck moving south at 10 m/s and the two vehicles stick together. What is the velocity of the truck immediately after the collision?

19. A white ball moving to the right hits a black ball that is at rest. After the collision, the black ball moves to the right and the white ball rebounds to the left. Which statement best describes the difference between the balls?
    A. The white ball is more massive than the black ball.
    B. The black ball is more massive than the white ball.
    C. The white ball is more elastic than the black ball.
    D. The black ball is more elastic than the white ball.

20. A 5 kg ball traveling at a speed of v hits and sticks to an identical ball at rest. After the collision, what will be the speed of the balls?
    A. $2v$
    B. $4v$
    C. $v/2$
    D. $v/4$

21. Which of the following is not conserved when a steel ball collides with another steel ball that is twice the mass of the first ball.
    A. Mass
    B. Kinetic energy
    C. Linear momentum
    D. Angular momentum

22. A white ball moving to the right hits a black ball that is at rest. After the collision, the black ball moves to the right and the white ball rebounds to the left. The speed with which the white ball rebounds is:
    A. Less than the speed of the black ball.
    B. Less than its speed before the collision.
    C. Greater than its speed before the collision.
    D. Equal to the speed of the black ball.

23. A 2 m long rod of negligible mass has a 40 N block attached to one end and it has a fulcrum located 1.2 meters from this block. How much mass must be placed on the other end if the rod is to be in equilibrium?

Answers to Questions and Problems

1. The new velocity is 2 times the old so the new kinetic energy is given by:
   $K_{new} = \frac{1}{2} m (2v)^2 = \frac{1}{2} m \, 4v^2 = 4(\frac{1}{2} m \, v^2)$ The answer is D.

2. The kinetic energy will be less when the projectile returns to the elevation at which it was launched because some of the mechanical energy was lost due to friction.

3. A. Kinetic energy decreases as the ball rises and increases as it falls.
   B. Gravitational potential energy increases as the ball rises and decreases as it falls.
   C. The momentum is the mass of the ball multiplied by its velocity vector. The velocity vector changes both in direction and magnitude.
   D. In the absence of air friction, the horizontal speed of a projectile does not change.

4. Mechanical energy is conserved, so the gravitational potential energy when the ball is dropped (mgh) must equal the kinetic energy at the ground ($\frac{1}{2} mv^2$). Solving for v we have: $v = \sqrt{2gh} = \sqrt{40}$ m/s. You should be able to guess that the square root of 40 is a little larger than 6. The answer is C.

5. The *magnitude* of the work done by friction is the force of friction times the distance traveled. (The work done by friction is always a negative quantity since the force of friction is opposite to the displacement.) The distance did not change, and the force of friction is equal to the coefficient of kinetic friction times the normal force. Neither depend on the speed of the object. The answer is C.

6. Kinetic energy is converted into gravitational potential energy and heat (because of the air friction). The "heat" is really thermal energy which shows up as an increase in the temperature of the projectile and the air it passes through.

7. The work done by the person is: $W = (40 \text{ N})(10 \text{ m}) = 400 \text{ J}$
   The power is: $P = W/t = (400 \text{ J})/(5 \text{ s}) = 80 \text{ W}$

8. Energy is conserved, so the initial energy of the projectile as it is launched (900 J of kinetic energy) is equal to the energy at the top of its trajectory. All the energy at the top is gravitational potential energy (mgh) since it was fired straight up and will continue up until it momentarily stops. Therefore: $mgh = 900$ J, and $h = (900 \text{ J})/(2 \text{ kg})(10 \text{ m/s}^2) = 45$ m
   (Using $g=9.8$ m/s^2 we would have gotten 46 m.)

9. The gain in gravitational potential energy of the box is given by:
$mgh = (50 \text{ kg})(10 \text{ m/s}^2)(2 \text{ m}) = 1000 \text{ J}$. If no energy is used to overcome friction, this is the work that the force F must do to raise the box. All the other information is extraneous. The answer is B.

10. The maximum height of the ball on the rebound can be found using $h = \frac{1}{2}gt^2$. This height is: $\frac{1}{2}(10 \text{ m/s}^2)(1/2 \text{ s})^2 = 1.25 \text{ m}$.
Since the potential energy is directly proportional to the height, the percentage of potential energy lost is equal to the percentage of height lost. The ball lost 0.75 m out of the original 2.0 m or $0.75/2 = 3/8$ or about 38%.

11. The mass does not affect the maximum height. Using the conservation of mechanical energy. The initial kinetic energy is equal to the potential energy at the top of the path (where the kinetic energy becomes zero.) Therefore: $\frac{1}{2}mv^2 = mgh$ and $h = v^2/2g = (30 \text{ m/s})^2/2(10 \text{ m/s}^2) = 45 \text{ m}$

12. The spring constant is the force that causes a stretch divided by the stretch. When an additional 0.2 kg was added to the spring, the spring stretched 0.005 m. The spring constant is:
$k = F/x = mg/x = (0.2 \text{ kg})(10 \text{ m/s}^2)/(0.005 \text{ m}) = 400 \text{ N/m}$

13. Power is the energy divided by the time. Therefore, the power is:
(150 megawatt-years)/(30 years) = 5 MW. The answer is D. The other three answers do not even have the correct units.

14. Since your momentum must be changed by a certain amount, no matter how much your shoes cost, the impulse ($\overline{F} \Delta t$) they give to you will be the same. However, shoes with a lot of cushioning will squash more than your cheap thin soled shoes. This squash, increases the *time* of the impulse and therefore decreases the force necessary to supply the needed impulse.

15. Momentum is conserved in this perfectly inelastic collision, but if the final mass is very large, its speed will be very small (approximately zero), and the kinetic energy will be extremely small (approximately zero). Therefore, nearly all the kinetic energy will be lost in the collision.

16. The flying pieces carry away some of the kinetic energy of the car, leaving less kinetic energy to be absorbed in the collision.

17. When a car crashes, it is given an impulse ($\overline{F} \Delta t$) by the object it hit. If parts of the car collapse, the time of the impulse is increased and the force is decreased accordingly.

18. The car has a momentum of 20,000 kg·m/s north and the truck has a momentum of 15,000 kg·m/s south. The net momentum before the collision is the *vector* sum of these two which is 5000 kg·m/s north. The momentum after the collision must also be 5000 kg·m/s north. The velocity afterwards is the momentum divided by the total mass which is: (5000 kg·m/s)/(2500 kg) = 2 m/s

19. The answer is B. Hopefully we knew this from being observant. If a small mass ball hits a large mass ball at rest, the small ball will rebound in the opposite direction and the large ball will move in the direction of the impulse imparted to it. If a large mass ball hits a small mass ball at rest, both balls will move in the same direction after the collision, but the small ball will have a larger speed. A more detailed analysis follows:

   Let m be the mass of the white ball and v be its velocity *before* the collision. Let M be the mass of the black ball and V be its velocity *after* the collision. The vector momentum must be conserved. Let momenta to the right be positive and momentum to the left be negative. Since the white ball has negative momentum (to the left) after the collision, the black ball must have more positive momentum after the collision than the white ball had before the collision. Although the white ball lost some kinetic energy, the black ball must have ended up with less kinetic energy than the white ball started with. Therefore, after the collision, the black ball had more momentum but less kinetic energy than the white ball had initially. In equation form we can write these statements as $MV > mv$ and $\frac{1}{2}MV^2 < \frac{1}{2}mv^2$ Solving the first equation for v we find $MV/m > v$ and substituting into the second equation we get $MV^2 < mM^2V^2/m^2$, which becomes $1 < M/m$ or $m < M$.

20. Obviously the speed is less so A and B can be eliminated immediately. Momentum is conserved. Before the collision the momentum was (5 kg)v and after the collision it was (10 kg)V where V is the final speed of the balls. Therefore, equating these momenta we find: $V = v/2$ The answer is C.

21. Some kinetic energy is always lost in a real collision. The answer is B.

22. The black ball carried away some of the kinetic energy the white ball had before the collision, leaving the white ball with less kinetic energy after the collision. (Some kinetic energy was also lost in the collision.) Therefore, the speed of the white ball must be less after the collision. The answer is B.

23. Draw a picture. Let F be the weight of the unknown mass which is located 0.8 m from the fulcrum. The net torque on the rod must be zero, (or the clockwise torques must equal the counterclockwise torques). Equating the torques we have: (40 N)(1.2 m) = F(0.8 m) Solving for F we find: $F = 60$ N Therefore, the mass must be: $m = F/g = (60$ N$)/(10$ m/s$^2) = 6$ kg

5 Fluids and Thermodynamics

5.1 Fluids

Matter generally comes in three phases: solids, liquids, and gases. Liquids and gases are called **fluids**, which are substances that do not maintain a definite shape. A solid has a definite shape and volume.

A characteristic property of a pure substance is its **density**. The density of an object (usually represented by the symbol ρ) is defined to be the mass (m) of the object divided by its volume (V).

$$\rho = \frac{m}{V} \qquad\qquad \textit{density}$$

The units of density are kilograms per cubic meter, but instead of stating the density of a substance, we often compare its density to the density of water. The ratio of the density of a substance to the density of water is a *dimensionless* quantity called the **specific gravity** of the substance.

An important concept when describing fluids is pressure. For a force applied *perpendicular* to an area of the fluid, the **pressure** is defined to be the force divided by the area.

$$P = \frac{F}{A} \qquad\qquad \textit{pressure}$$

The units of pressure are newtons per square meter, and one N/m^2 is defined to be one pascal (abbreviated Pa). The pressure at the bottom of the Earth's atmosphere is about 1×10^5 Pa (about 15 lb/in^2). Since we live at the bottom of a fluid that constantly supplies a pressure of about 15 lb/in^2, most pressure gauges register only the value of the pressure that is *above* atmospheric pressure. The pressure above atmospheric pressure is called the **gauge pressure**. To get the total or **absolute pressure**, atmospheric pressure must be added to the gauge pressure.

Fluids that are not moving or circulating appreciably are said to be static. **Hydrostatics** is the study of fluids that are not moving. The **hydrostatic pressure** in a fluid increases with depth, and if the fluid is static or nearly so, the pressure P at a depth h in the fluid is equal to the pressure on the surface of the fluid plus the weight of the fluid above a unit area at that depth. This statement can be written in equation form as:

$$P = P_o + \rho g h \qquad\qquad \textit{hydrostatic pressure}$$

Where ρ is the density of the fluid (ρg is the weight per unit volume) and P_o is the pressure on the top of the fluid. If the fluid is exposed to the air, then the pressure on the top of the fluid is atmospheric pressure, and the gauge pressure in the fluid is just $\rho g h$.

Example 5.1

In a fluid with a density of 1.0×10^3 kg/m³, find the difference in pressure between two points if one of the points is 30 cm above the other.

The difference in pressure is given by $\rho g h$ so the difference is:

$$\rho g h = (1 \times 10^3 \text{ kg/m}^3)(10 \text{ m/s}^2)(0.3 \text{ m}) = 3 \times 10^3 \text{ N/m}^2$$

Archimedes' Principle

Since the pressure in a fluid increases with depth, an object submerged in a fluid will experience a greater pressure on its bottom than on its top. This difference in pressure causes an upward force on the object that is known as the **buoyant force** exerted by the fluid.

An important principle in connection with fluids is **Archimedes' Principle**, which states that the buoyant force on a body immersed in a fluid is equal to the weight of the fluid displaced by the object. This principle says that a floating object will sink down into the fluid until the fluid that is displaced is equal to the weight of the object. Therefore, that fraction of the object that is submerged will be the ratio of the average density of the floating object to the density of the fluid. In equation form this statement can be written as:

$$f = \frac{\rho}{\rho_f}$$

Where f is the fraction of the object that is submerged, ρ is the average density of the object, and ρ_f is the density of the fluid.

Example 5.2

A uniform block of wood with dimensions of 25 cm by 15 cm by 4 cm and a density of 600 kg/m³ is placed in a bucket of water whose density is 1000 kg/m³. How deep into the water will the block sink?

The wood will float with part of the 4 cm high side down into the water. The fraction of the volume that is submerged is equal to the ratio of the densities. Therefore, the submerged volume is:

$(600 \text{ kg/m}^3)/(1000 \text{ kg/m}^3) = 0.6$

The depth of the block will be: $(0.6)(4 \text{ cm}) = 2.4 \text{ cm}$

Fluids in Motion

The study of *moving* fluids is called **hydrodynamics**. As a fluid moves through a pipe of varying size, the *mass* of fluid passing by any point per second must remain constant. This statement can be written as an equation that is commonly referred to as the **equation of continuity**. If we consider two points along the pipe we can write:

$$\rho_1 A_1 v_1 = \rho_2 A_2 v_2$$

or

$$\rho A v = \text{constant} \qquad\qquad \textit{equation of continuity}$$

Where ρ is the fluid's density, A is the pipe's cross sectional area, and v is the velocity of the fluid at the given point. The density of a fluid will increase as the pressure increases, but under normal conditions liquids are nearly incompressible, and often the density of a gas can be *approximated* as a constant. Under these conditions, the equation of continuity becomes:

$$A v = \text{constant}$$

Example 5.3

A portion of an artery is constricted to one-half its normal diameter. How does the speed of the blood in this section compare to the speed in the normal part of the artery?

Since the density of blood does not change appreciably, the continuity equation requires that the velocity times the cross sectional area must be constant along the artery. The cross sectional area is proportional to the square of the diameter so the constricted region has one-fourth the cross sectional area of the normal artery. Therefore, the speed in the restricted area must be 4 times the speed in the normal region.

A important principle in fluid flow is **Bernoulli's principle**. It states that where the velocity of a fluid is *high* the pressure is *low*, and were the velocity is *low* the pressure is *high*. When we first hear this statement, it seems to contradict our common sense. However, it does make sense if we consider a fluid flowing through a fat pipe that is connected to a thinner pipe. We know that the fluid moves faster through the thin pipe, and as the fluid enters the thin section of pipe it has to speed up. In order to speed up, there must be a *net* force on the fluid in the direction it is moving. However, the excess force must come from the *fat* section of pipe and, therefore, the pressure in the fat section must he *higher* than the pressure in the thin section.

If we allow for the fact that the elevation of the pipe may also change, Bernoulli's equation becomes:

$$P + \tfrac{1}{2}\rho v^2 + \rho g y = \text{constant} \qquad \textit{Bernoulli's equation}$$

Where P is the pressure in the fluid, ρ is the density of the fluid, v is the velocity of the fluid, and y is the height of the fluid. Bernoulli's equation is just a statement of the conservation of energy for a fluid. The second and third terms are easily recognized as the kinetic energy per unit volume and the gravitational potential energy per unit volume of the fluid. Bernoulli's equation is generally used by evaluating the three terms at one point and equating them to the three terms evaluated at another point.

$$P_1 + \tfrac{1}{2}\rho v_1^2 + \rho g y_1 = P_2 + \tfrac{1}{2}\rho v_2^2 + \rho g y_2 \qquad \textit{Bernoulli's equation}$$

Example 5.4

A water tank filled with water to a depth of 10 meters springs a leak near the bottom of the tank. Assuming that atmospheric pressure is 1×10^5 N/m^2, find the speed of the water leaving the hole.

We apply Bernoulli's equation to the water near the top of the tank and to the water coming out of the hole. The value of h in Bernoulli's equation is 10 m, the difference in elevation of the two points of interest. The pressure at both places is the same (both are at atmospheric pressure). Atmospheric pressure cancels out on both sides of the equation, so its value is not important. The water at the top of the tank is *not* moving so Bernoulli's equation becomes: $\rho g h = \tfrac{1}{2}\rho v^2$

If we cancel out the density we obtain:

$v = \sqrt{2gh} = \sqrt{2(10 \text{ m/s}^2)(10 \text{ m})} = 14$ m/s

5.2 Viscosity, Turbulence, and Surface Tension

All real fluids have internal friction that is called **viscosity**. As a fluid moves through a pipe or along a surface, there is a certain amount of friction between the fluid and the pipe or surface. This friction acts to resist the flow of the fluid. Similarly, when an object moves through a fluid, the fluid flowing around the object exerts a resistive force or **drag force** on the moving object. This drag force increases as the velocity of the object increases. As an object falls through the air and gains speed, the drag force increases until it balances out the pull of gravity. Beyond this point, the particle no longer *accelerates*, but continues to fall at a constant speed (called its terminal velocity).

Anyone who has stood by a river has noticed that the water does not flow uniformly down stream, but it swirls and turns violently in many places. We say the flow is **turbulent**. Little whirl pools (called Eddy currents) are common near rocks and bends in the river. Even fluids flowing through smooth tubes can have turbulent flow under certain conditions. This turbulent flow is caused by the viscosity or internal resistance of the fluid. Viscosity of a liquid is caused by the friction between the molecules. The flow of fluids through most pipes is turbulent flow as is the flow of blood through our circulatory system. In gases, turbulent flow is caused by collisions between the molecules of the gas.

The molecules in a liquid are attracted by their surrounding neighbors. The net force on molecules *inside* the liquid is zero since they are completely surrounded by attracting neighbors, but the molecules on the *surface* of the liquid are attracted to the rest of the liquid. This inward attractive force causes the surface to contract and resist being distorted. This resistance to distortion is a property of a liquid we call **surface tension**. Surface tension acts like a membrane on the surface of a liquid. Surface tension causes a raindrop to be pulled into a spherical shape as it falls. It also allows small insects like the water strider to walk on the surface of the water and it will float a needle on the surface of a glass of water. The membrane-like nature of the water's surface will be depressed slightly by light objects and will support the weight of certain objects.

5.3 Heat and Thermodynamics

Temperature

In a qualitative sense, the temperature of an object is a measure of how hot or cold the object is. There are three scales in common use to measure temperature. They are the **Celsius** or **centigrade** scale, the **Fahrenheit** scale (in the United States), and the absolute or **Kelvin** scale. The Kelvin scale is the most important for scientific work because the zero point on this scale is the lowest possible temperature, the temperature at which an object is in its lowest possible energy state. The zero point on the Celsius scale (0°C) is the freezing point of water which is 273 on the Kelvin scale (written 273 K).

Thermal Expansion

The atoms and molecules in a solid vibrate about their equilibrium position. As the temperature rises, the vibrations become more vigorous and the space occupied by each particle increases, causing the solid to expand. Experiments indicate that a solid with an original length L_0 will expand an amount ΔL as the temperature changes by an amount ΔT. These quantities are related by the expression:

$$\Delta L = \alpha L_0 \, \Delta T \qquad\qquad\qquad\qquad \textit{linear expansion}$$

where the proportionality constant α is called the **coefficient of linear expansion**, a quantity measured in units of $(C^\circ)^{-1}$. A similar expression can be written for the volume expansion of an object:

$$\Delta V = 3\alpha V_0 \, \Delta T = \beta V_0 \, \Delta T \qquad\qquad\qquad \textit{volume expansion}$$

where β, the coefficient of volume expansion, is 3α for most solids (solids that expand symmetrically in all *three* directions).

Gas Laws

The relationship between the pressure (P), volume (V), and the absolute or Kelvin temperature (T) of a sample of gas is described by the laws of Boyle, Charles, and Gay-Lussac. They are summarized below:

Boyle's law: $\qquad\qquad\qquad V \propto \dfrac{1}{P} \qquad\qquad$ *(constant T)*

Charles's law: $\qquad\qquad\quad V \propto T \qquad\qquad$ *(constant P)*

Gay-Lussac's law: $\qquad\quad\ P \propto T \qquad\qquad$ *(constant V)*

These three laws can be combined into one equation that describes how the various quantities are related. This equation is known as the **ideal gas law** or the **equation of state** for an ideal gas. Real gases do not follow the equation precisely, but it describes the behavior of most gases very nicely. It is given by:

$$PV = nRT \qquad\qquad\qquad\qquad\qquad \textit{ideal gas law}$$

Where n is the number of moles (abbreviated mol) in the sample, and R is a constant called the **universal gas constant**. The universal gas constant has a value of:

$$R = 8.315 \text{ J/(mol·K)}$$ *universal gas constant*

A **mole** of a substance is an a quantity that contains a specific *number* of particles (atoms or molecules). That number is called **Avogadro's number**, and is the number of atoms in 12 grams of carbon 12. Carbon 12 is the isotope of carbon that contains 12 nucleons (6 protons and 6 neutrons). Avogadro's number is:

$$N_A = 6.02 \times 10^{23} \text{ particles/mol}$$ *Avogadro's number*

For a specific quantity of gas, a change in either the temperature, pressure, or volume of the gas can be easily calculated. The equation relating the initial values of these quantities to the final values can be expressed as:

$$\frac{P_i V_i}{T_i} = \frac{P_f V_f}{T_f}$$

The pressure of a gas is the result of particle collisions (collisions by atoms and/or molecules). As the particles of a gas collide with the walls of its container, they exert a force on the walls of the container. This force on the area of the walls is the pressure of the gas. As the temperature of a gas increases, the average speed of the gas particles increases. The absolute temperature of a gas is a direct measure of the *average* translational kinetic energy of a particle in the gas. This statement is given by the expression:

$$\left(\tfrac{1}{2} m v^2\right)_{ave} = \tfrac{3}{2} kT$$

where $k = R/N_A$ is the universal gas constant divided by Avogadro's number. It is called **Boltzmann's constant** and has the value: $k = 1.38 \times 10^{-23}$ J/K

Example 5.5

A 2 liter container of gas at a temperature of 300 K and a pressure of one atmosphere is compressed to half its original volume. If its pressure rises to 3 atmospheres, what is its new temperature?

We can use the expression: $\dfrac{P_i V_i}{T_i} = \dfrac{P_f V_f}{T_f}$

Since we are working with ratios, we do not need to change the volume or pressure to SI units, however, the temperature must be the Kelvin temperature. Solving for the final temperature and substituting in the given quantities we have:

$$T_f = T_i \frac{P_f V_f}{P_i V_i} = (300 \text{ K})(3 \text{ atm})(1 \text{ liter})/(1 \text{ atm})(2 \text{ liter}) = 450 \text{ K}$$

Example 5.6

The molecular mass of an oxygen molecule is about 32 and the molecular mass of a nitrogen molecule is about 28. In a room where the temperature is 68°F which of the following is true?
A.   The average translational kinetic energy of the oxygen molecule is 7/8 times the average translational kinetic energy of the nitrogen molecules.
B.   The average translational kinetic energy of the oxygen molecules is 8/7 times the average translational kinetic energy of the nitrogen molecules.
C.   The average translational kinetic energy of the oxygen molecules is the same as the average translational kinetic energy of the nitrogen molecules.
D.   The average speed of the oxygen molecules is greater than the average speed of the nitrogen molecules.

The temperature is a measure of the average translational kinetic energy. The answer is C. Since they have the *same* average kinetic energy but the oxygen molecule is more massive, the average *speed* of the oxygen molecules is *less* than the average speed of the nitrogen molecules.

Heat

Heat is the name given to energy that is transferred from one object to another because of a difference in temperature. Heat is generally transferred by one of three means: conduction, convection, or radiation. **Conduction** is the transfer of *internal* energy from one location in an object to another location. This energy is microscopic vibrational energy that is transferred from particle to particle by collisions, and it is generally the important method of energy transfer in *solids*. Conduction is how the energy is transferred from the hot end of a rod to the cooler end.

Convection is the transfer of energy by *mass motions* within a fluid. It is often the most important means of energy transfer in *fluids*. As the air around the flame of a candle is warmed (gains energy from the candle), This more "energetic" air expands and the increased buoyant force on the warm air causes it to rise. The stream of rising warm air is called a convection current.

All objects are continually emitting and absorbing energy in the form of electromagnetic waves. Most of these electromagnetic waves will be infrared light waves if the temperature of the object is below a few hundred degrees Celsius. An object that is warmer than its surroundings will emit more energy than it absorbs. The transfer of energy (in the form of infrared light) from a warm object to its surroundings is called **radiation**. Objects with a temperature above about 700°C (about 1000 K) will begin to glow red as they radiate a significant fraction of visible light. The warmth you feel while sitting by a fire is energy (in the from of infrared light) that is being transferred from the fire and fire place bricks to you by radiation. Energy from the Sun (in the form of infrared, visible, and ultraviolet light) is transferred to Earth by radiation. This electromagnetic energy is often just called radiation.

Heat is a type of energy, and energy is measured in units of joules. However, heat is often measured in calories. One **calorie** (abbreviated cal) is defined to be the amount of energy that is needed to raise the temperature of one gram of water by one Celsius degree. (Actually from 14.5°C to 15.5°C since the amount of energy varies slightly with the temperature.) One calorie is about 4 joules. The conversion is:

1 cal = 4.18 J   *calorie*

The amount of heat (energy) need to change the temperature of one gram of a substance by $1C°$ is called the **specific heat** of the substance. The specific heat of water is 1 cal/g·C° = 1 kcal/kg·C°. The relationship between the heat added Q, the change in temperature ΔT, and the mass of the substance m is given by:

$Q = mc\ \Delta T$ *specific heat equation*

Where c is the specific heat of the substance. The value of the specific heat of water is one of the highest of any known substance. This high value helps explain why the temperature of the ocean does not change much with the seasons. The high specific heat of water makes it easier for our bodies to maintain a constant temperature (animal tissue is mostly water).

Example 5.7

A certain fluid has a specific heat of 3 J/g·°C. If 500 g of it is heated by a 1.2 W heater, how much will its temperature change in 100 seconds?

In 100 s the heater will put out (1.2 W)(100 s) = 120 J of heat. The specific heat is related to the heat added by the expression: $Q = mc\ \Delta T$. Solving for ΔT we obtain:

$\Delta T = Q / mc = (120\ \text{J})/(500\ \text{g})(3\ \text{J/g·°C}) = (120/1500)\ °C = 0.08°C$

Internal Energy

The total energy stored by the particles that comprise an object is called the **internal energy** of the object. This energy may be in several forms. The particles may have rotational kinetic energy, translational kinetic energy, and various forms of potential energy. Although internal energy is sometimes called **thermal energy**, it is technically incorrect to call it heat. (Remember that heat is energy that is *transferred* from one object to another.) As energy (heat) is absorbed by an object, the object's internal energy increases. Unless the object is changing its phase, this increase in internal energy will cause the temperature of the object to increase.

A **change in phase** is a change from a solid to a liquid, or a change from a liquid to a gas, or the reverse of either of these. A change of phase requires a change in internal energy, but a change in the temperature of the substance does *not* accompany this transfer of energy. The amount of energy required to change 1 kg of a solid to a liquid is called the **heat of fusion** or **latent heat of fusion**. The heat of fusion of water is about 79.7 cal/g = 79.7 kcal/kg. It takes 79.7 cal to melt 1 g of ice at 0°C. (To freeze 1 g of water at 0°C, 79.7 cal of energy would have to be removed from the water.)

The amount of energy required to change 1 kg of a liquid to a gas is called the **heat of vaporization** or **latent heat of vaporization**. The heat of vaporization of water is 539 cal/g = 539 kcal/kg. It takes 539 calories to convert 1 g of boiling water into steam. To convert the steam back to water requires the removal of 539 calories of energy.

First Law of Thermodynamics

The **first law of thermodynamics** is just a statement of the *conservation of energy* as applied to a system (like a sample of gas or a glass of water). If we add some energy to a particular system, some of that energy could be used to change the internal energy of the system and some of the energy could be used by the system to do work. This statement can be written as:

$$Q = \Delta U + W \qquad \qquad \textit{first law of thermodynamics}$$

Where Q is the energy added to the system, W is the work done *by* the system, and ΔU is the *change* in internal energy of the system.

Work Done by a Gas

If a sample of gas with a pressure P is confined to a container and one wall of the container (with an area of A) moves a distance Δx, the volume of the gas will increase by an amount $\Delta V = A \, \Delta x$. The pressure of the gas is the force per unit area ($P = F/A$), and the work done by the gas is:

$$W = F \, \Delta x = F \, \Delta V / A = (F/A) \, \Delta V = P \, \Delta V$$

The work done by a gas is generally written as:

$$W = P \, \Delta V \qquad \qquad \textit{work done by a gas}$$

If the volume *increases* (ΔV is positive) the work done is *positive*, but if the volume *decreases* (ΔV is negative) the work done by the gas will be *negative*. When a sample of gas does some work, the energy to do the work must come from somewhere. If no energy is added or removed from the gas, the energy must come from the *internal* energy of the gas, and the temperature of the gas will decrease.

If the gas does *negative* work (and no energy is added or removed), the internal energy of the gas will *increase*. This is why a sample of gas gets hotter when it is compressed. The gas does negative work when it is compressed, and if the compression happens fairly quickly there is not enough time for a significant amount of heat to enter or leave the gas. If a sample of gas expands or contracts quickly there generally will not be enough time for much heat to be added or removed from the sample, and Q in the first law will be zero. (Solids and liquids do not expand or contract appreciably, so the work they do can generally be ignored.)

Second Law of Thermodynamics

The first law of thermodynamics states that energy is conserved, but we know there are certain processes that seem to be prohibited by nature even though energy is conserved. For example, if warm water is placed in a glass, we all know the water will cool off and the room will get a little warmer. However, the first law of thermodynamics does not prohibit the water from getting warmer as the room cools off. The **second law of thermodynamics** tells us which processes are permitted in nature and which ones are not. There are many equivalent ways to state the second law of thermodynamics and several are given below:

> Heat will not flow *spontaneously* from a cold object to a hot object.

> No device can transform a given amount of heat *completely* into work.

> It is impossible to create a perpetual motion machine.

> Natural processes proceed toward a state of greater *disorder*.

The second law of thermodynamics says that a heat engine cannot convert heat completely into work. A **heat engine** is a device that uses heat from a high temperature reservoir (such as steam or hot air) and uses a portion of this heat to do work. The unused heat is dumped into a cooler temperature reservoir (the exhaust).

The amount of work that a heat engine can produce depends on the temperatures *between* which it operates. For example, a steam turbine is turned by high temperature steam. As the steam passes through the turbine blades, the steam gives some of its energy to the turbine and passes out of the turbine at a lower temperature. The theoretical **efficiency** of such a heat engine is determined by the temperatures of the reservoirs. If T_H is the high temperature (measured in degrees Kelvin) and T_L is the low temperature between which the heat energy is operated, the maximum efficiency of the heat engine is:

$$e = 1 - \frac{T_L}{T_H}$$
efficiency of heat engine

The efficiency approaches one (100% efficient) as the high temperature increases or as the low temperature approaches zero. However, the low temperature is generally the temperature of the environment (about 300K), and if the high temperature gets too high it will begin to melt the engine.

A **refrigerator** or **heat pump** is a heat engine running in reverse. The machine does work to remove heat from a low temperature reservoir (the inside of the refrigerator) and deposits the heat into a higher temperature reservoir (the air outside of the refrigerator). The amount of energy dumped into the high temperature reservoir is equal to the energy removed from the low temperature reservoir *plus* the work done by the refrigerator to remove this heat. A perfect refrigerator (which is prohibited by the second law of thermodynamics) would require no work to remove heat from the lower temperature reservoir.

The quantity that measures the disorder of a system is called the **entropy**. The most disordered states are those states that are the most probable. There are more ways to assemble a disordered state than there are to assemble an ordered state and, therefore, one of the disordered states are more likely to be the actual state of the system. The second law of thermodynamics says that the entropy of a closed system cannot decrease.

Questions and Problems

1. An object that is just slightly more dense than the water at the surface of the ocean will:
   A. float on the surface.
   B. sink until it reaches the bottom.
   C. sink until its average density matches the density of the fluid.
   D. sink until its average density is slightly less than the surrounding fluid.

2. A ball with an average density of 8×10^2 kg/m^3 is floating in a pool of water (with a density of 1×10^3 kg/m^3). What fraction of the ball is above the water?

3. If a warm wire is immersed in a beaker of oil, the oil around the wire will be heated. How will the density and position of heated oil change?
   A. The density of the heated oil will increase and it will rise.
   B. The density of the heated oil will not change.
   C. The density of the heated oil will increase and sink.
   D. The density of the heated oil will decrease and rise.

4. Seat belts are designed to restrain a person in the event of a collision. It seems like thinner belts would be more comfortable, why don't they make narrower belts?

5. Explain how Bernoulli's equation applied to the wings of an airplane keeps the airplane in the air.

6. When flying at a given speed, how does the lift on airplane wing near sea level compare to the lift on the identical wing at an elevation of 30,000 feet?

7. If a sink is filled to the top with water, a force of 25 N is required to pull the plug. If the sink is filled to the top with motor oil whose density is 800 kg/m, how much force will be required to pull the plug?

8. How long will it take a 1.0 W heater to change the temperature of 1 kg of water by 0.1°C? (The specific heat of water is 4.2×10^3 J/kg·°C.

9. The force needed to pull the plug in a fluid filled sink is directly proportional to:
   A. the diameter of the plug
   B. the viscosity of the fluid
   C. the area of the sink
   D. the depth of the fluid

10. When your skin gets cut, is the cause the net force applied to it or the pressure.

11. The velocity of sound in air is given by the expression: $v = \sqrt{\dfrac{1.4\,RT}{M}}$

Where R is the universal gas constant, T is the temperature in degrees Kelvin, and M is the molar mass of the gas. If the temperature increases by 21%, by what percent does the speed of sound change?

12. Which of the following expressions gives the molecular mass (or weight) of a gas from its density, D?
A. DP/RT
B. nRT/DP
C. DRT/P
D. P/DRT

13. Describe the energy transformation that occurs as steam pushes the piston of a steam engine.

14. As a closed container of water is heated, the pressure increases. There are two reasons for the increase in pressure. What are they?

15. A heat engine running between the temperatures of 300 K and 500 K. What is the theoretical efficiency of the engine and what could be done to increase the efficiency?

16. If we need a heat engine to do 1000 J of work, how much heat would be required to do produce this work?
A. 500 J
B. 750 J
C. 1000 J
D. 4000 J

17. When you come out of a swimming pool on a hot windy day, you often get chilled. Why does this happen?

18. Explain why opening the refrigerator door on a hot day will only cool you off temporarily.

19. Consider two identical tanks, one filled with water and the other filled with oil. The force required to pull the plug at the bottom of the oil filled tank will be:
A. greater because the oil is more viscous.'
B. less because the oil is less viscous.
C. greater because the oil is more dense.
D. less because the oil is less dense.

Answers to Questions and Problems

1. When the density of the surrounding fluid matches the average density of the object, the weight of the object will equal the buoyant force (the weight of the fluid displaced) and the net force will be zero. The **answer is C**.

2. The fraction below the surface is just the ratio of the densities which is:
$(8 \times 10^2 \text{ kg/m}^3)/(1 \times 10^3 \text{ kg/m}^3) = 0.8 = 4/5$ Therefore the fraction above the surface is 1/5.

3. Most substances expand when they are heated. Therefore, their density decreases and their buoyant force increases, causing them to rise. The answer is D.

4. To restrain a person in a collision, the belt must apply a force of a certain magnitude over a short period of time (an impulse of a given size). A wide belt allows this force to be spread over a larger area and, therefore, results in a smaller *pressure*. We react to pressures not forces. Large forces distributed over large areas create less pressure and are less painful than large forces distributed over smaller areas (high pressures).

5. The wings are designed so the path over the top of the wing is further than the path along the bottom. Therefore, the air must move faster over the top of the wing. The pressure on top of the wing (where the air is moving faster) is less than the pressure on the bottom of the wing. It is this *difference* in pressure that supplies the airplane's lift. To remain in level flight, the lift must be equal to the weight of the plane.

6. The lift is due to the difference in pressure on the top and bottom of the wings ($P_{bottom} - P_{top}$). Bernoulli's equation tells us that this difference is:
$\frac{1}{2}\rho(v^2_{top} - v^2_{bottom})$ (The height difference between the top and bottom of the wing is negligible.) The pressure difference is proportional to the density of the air. At 30,000 ft, the density of air is less, so the lift will be less. Just because an airplane can fly at an elevation of 3000 ft does not guarantee that the plane can fly at 30,000 ft.

7. The pressure at a given depth is given by: $\rho g h$ Since the pressure is proportional to the density of the fluid and the force on the plug is proportional to the pressure, the force will be reduced by:
$(800 \text{ kg/m}^3)/(1000 \text{ kg/m}^3) = 0.8$
Therefore the force required will be: $(0.8)(25 \text{ N}) = 20 \text{ N}$

8. Since 4.2×10^3 J will increase the temperature of 1 kg of water by 1°C, we need 4.2×10^2 J to increase 1 kg of water by 0.1°C. The 1 W heater puts

71

out 1 J of heat each second, so we must run the heater for 4.2×10^2 s. (About (420/60) minutes = 7 minutes.)

9. A. The force is proportional to the *area* of the plug (the square of the diameter.
    B. The viscosity will have a small effect on the force, but it is not directly proportional to the viscosity.
    C. The area of the sink has no effect on the force to pull the plug.
    D. The force needed on the plug is proportional to the pressure, and the depth is proportional to the pressure, so the force is proportional to the depth. The answer is D.

10. Pressure (not force) is responsible for cuts. A sharp needle can easily pierce the skin when a small force is applied, but the same force may not cause a dull needle (one with a larger surface area resulting in a smaller pressure) to pierce the skin. A given force can produce very high pressures when applied to a very tiny area. Sharp knives cut easier for the same reason.

11. If v is the original velocity of sound and T is the original temperature, the new temperature is changed to $1.21\,T$ (increased by 21%). Since the velocity is proportional to the square root of the temperature, the velocity changes to $\sqrt{1.21}\,v = 1.1\,v$. Therefore the velocity is increased by 10%.

12. Attack this problem with dimensional analysis. The molecular weight has units of mass/mole and since the density (D) has units of mass/volume, the answer must have a D in the numerator. This knowledge eliminates answers B and D. In addition, R has units of J/mole·K so the correct answer must have an R in the numerator (an R in the denominator along with D would give units of mass·mole not mass/mole.) The correct answer must be C. As a check on the units, DRT/P has units of:
$(\text{kg/m}^3)(\text{J/mole·K})(\text{K})/(\text{N/m}^2) = (\text{kg/m}^3)(\text{N·m/mole})/(\text{N/m}^2) = (\text{kg/mole})$

13. The internal energy of the gas (or heat from the gas) is used to produce mechanical energy of the piston.

14. As the temperature rises, the number of water molecules in the vapor above the water increases. Also, the increase in temperature causes an increase in the pressure of the vapor.

15. The efficiency is: $e = 1 - \dfrac{300 \text{ K}}{500 \text{ K}} = 1 - 0.6 = 0.4$
    The efficiency could be increased by increasing the higher temperature.

16. Much more than 1000 J since the efficiency of a heat engine can never be 100%. Only D satisfies this requirement.

17. The wind causes the water on you body to evaporate, but it takes a great deal of energy to change water to a vapor and this energy must come from the surrounding environment. Much of the energy comes from your body, which cools you off.

18. With the door open the refrigerator will soon begin to run continuously. Since the heat removed from inside the refrigerator is dumped into the room along with the energy required to remove the heat, in the long run the room will get hotter with the refrigerator running than it would with the refrigerator off or running with the door closed.

19. The oil is more viscous but this has little to do with the force required to pull the plug. The difference is due to the higher pressure at the bottom of the denser liquid which is the water. The answer is D.

6 Vibrations and Waves

6.1 Vibrations

As we observe the natural world, we often encounter oscillating or vibrating objects. For example, a flagpole wiggles as the wind blows, a child on a swing oscillates back and forth, and the string on a violin vibrates as the bow is drawn across it. If you hit the wall with your fist, the wall and windows will vibrate, and your coffee table may vibrate as certain notes are played on your stereo. An object will oscillate if the object has a tendency to *return* to its original configuration when it is moved or distorted.

A simple oscillating system is a mass attached to a spring. The **equilibrium position** of the mass is its location when the net force on the mass is zero. (The position where the upward force of the spring balances the downward gravitational pull of Earth.) If the mass is moved so the spring is stretched or compressed and then the mass is released, the spring will exert a force on the mass that tends to *return* it to its equilibrium position. As the mass accelerates toward the equilibrium position, it gains speed and overshoots the equilibrium position. On the other side of the equilibrium position, the spring causes the mass to slow down and eventually stop. The motion continues as the mass oscillates about its equilibrium position. Of course, the motion eventually stops because of friction. The force (F) exerted on the mass by the spring is given by **Hook's law**:

$$F = -kx \qquad\qquad\qquad Hook's\ law$$

Where x is the displacement of the mass from its equilibrium position and k is the spring constant. The negative sign indicates that the force and displacement are always in the opposite directions (called a restoring force). If the motion of an oscillating system is caused by a restoring force that is directly proportional to the displacement (given by Hook's law), the motion is said to be **simple harmonic motion**. The **amplitude** of the motion is the *maximum displacement* of the mass (its maximum distance from the equilibrium position). For *any* restoring force, the motion will be simple harmonic if the amplitude is *small* enough. This amazing fact is true because if you take a small enough displacement, the force will be approximately proportional to the displacement.

The **period** (T) of the motion is the time for the mass to move through one complete cycle, which is one round trip. The **frequency** (f) is the number of cycles completed in one second. As can be deduced from the definitions (the period is the "seconds per cycle" and the frequency is the "cycles per second"), the frequency is the inverse of the period:

$$T = \frac{1}{f} \qquad\qquad\qquad period\text{-}frequency\ equation$$

75

Periods are measured in seconds and frequencies are generally expressed in hertz (abbreviated Hz). One hertz is equal to one cycle per second. The period of oscillation of a mass on a spring depends on the mass and the spring constant and is given by:

$$T = 2\pi\sqrt{\frac{m}{k}}$$ *period of mass on a spring*

Experience tells us that the period of vibration *increases* as the mass increases, and the period *decreases* as the spring constant increases. Short period vibrations are produced by stiff springs (large *k* values) and/or small masses.

Another common example of simple harmonic motion is a **simple pendulum**. A simple pendulum is made by suspending a mass (called the bob) from a string of negligible mass. The mass swings along the arc of a circle an equal distance on either side of its equilibrium position. The amplitude of the motion is the bob's maximum displacement from its equilibrium position. The motion will be simple harmonic if the amplitude is *small* compared to the length of the string. Small is a relative term, but if the arc is less than about 10°, the motion can be considered to be simple harmonic. Any type of oscillatory motion will be simple harmonic *if* the amplitude of the motion is small enough.

Since an object's acceleration due to gravity does *not* depend on the mass of the object, the period of a simple pendulum does not depend on the mass of the bob. A pendulum's period only depends on the length of the string (*L*) and the acceleration of gravity (*g*). It is given by:

$$T = 2\pi\sqrt{\frac{L}{g}}$$ *period of simple pendulum*

As expected, the period increases (takes longer to swing back and forth) as the length of the string increases, and the period decreases as the acceleration of gravity increases. (If the pendulum were in outer space were there is no gravity, the bob would not move, which is an infinite period.) A pendulum of a known length can be used to measure the acceleration of gravity to a very high degree of accuracy.

Example 6.1

A mass attached to a spring oscillates in simple harmonic motion. If the spring is replaced by a spring with a spring constant that is 44% larger, by what percent does the period increase?

The spring constant is increased to 1.44 times the old value. Since the period is proportional to the square root of the spring constant, the new period is equal to $\sqrt{1.44} = 1.2$ times the old value. Therefore, the period increases by 20%.

6.2 Waves

A wave is a *disturbance* that propagates from one point to another. **Mechanical waves** are disturbances that propagate through a *material* medium. Mechanical waves include waves traveling along a string, sound waves traveling in air (or in other substances), seismic waves traveling through the Earth, and water waves. Light is *not* a mechanical wave since it does not need a propagating medium.

A disturbance of short duration will produce a wave pulse, but if the wave is produced by a disturbance that is executing simple harmonic motion, the wave will be a continuous periodic wave called a **harmonic wave** or sinusoidal wave. For example, if you shake the end of a long rope in a regular manner, a harmonic wave will travel along the rope. If the motion of your hand is simple harmonic, then every piece of the string along its length will also execute simple harmonic motion.

If the particles in the medium oscillate in a direction that is *perpendicular* (or *transverse*) to the direction in which the disturbance is moving, the wave is said to be a **transverse wave**. A wave traveling along a string or rope is an example of a transverse wave. For example, as a wave travels from left to right along a rope, different parts of the rope oscillate up and down. If the particles in the medium oscillate *along* the direction in which the disturbance is traveling, the wave is called a **longitudinal wave**. Sound waves are longitudinal waves. If a sound wave is traveling to the right, the air molecules will be oscillating right and left.

A water wave appears to be a transverse wave, but it is actually more complicated since it is a disturbance that only occurs near the surface of the medium (the water). The particles of water move in circles (up and down as well as back and forth) as the wave travels along the surface of the water. Nevertheless, water waves are easy to visualize when discussing waves, and they are often used as an example of wave motion.

The high points on the wave are called **crests** and the low points are called **troughs**. In periodic waves, the wave pattern repeats itself at regular intervals. The length of this regular interval (for example, the distance from one crest to the next crest) is called the **wavelength**. The **frequency** of the wave is the number of wavelengths that pass a given point per unit of time (per second). Frequency is measured in cycles per second (or hertz).

The **wave velocity** is the speed with which a wave crest moves (or any other point on the wave pattern). The speed of a wave (v) is just the number of wavelengths that pass a point per unit time (the frequency f) multiplied by the length of a wavelength (λ). These three quantities are related by:

$$v = \lambda f \qquad\qquad\qquad\qquad\qquad \textit{wave velocity}$$

The speed of a wave depends on the physical properties (the density for example) of the medium in which it is traveling. The **amplitude** of a wave like a water wave is the height of a crest above the *average* level of the water (or the depth of a trough below the average level of the water). This definition is equivalent to saying that the amplitude of a wave is the maximum *displacement* of a particle from its equilibrium position. Since each particle in the medium is executing simple harmonic motion, the amplitude of the wave is just the amplitude of the simple harmonic motion of a particle of the medium. The frequency of the wave is just the frequency of the particle's simple harmonic motion.

When a wave arrives at a boundary that separates one medium from another (or a boundary that separates portions of the same medium with different physical properties), part of the wave bounces off the boundary (the **reflected wave**) and part of the wave passes into the other medium (the **refracted wave**). The refracted wave is so named because its direction is generally changed as it passes into the second medium. This change in direction is due to the fact that the wave's speed *changes* as the wave enters a different medium.

Example 6.2

A mechanical wave has an amplitude of 4.0 N/m^2, a frequency of 2000 Hz, a phase angle of $\pi/2$, a wavelength of 3 m, and an intensity of 8 x 10 W/m^2. Find the speed of the wave.

The speed is equal to the wavelength times the frequency, the other information is not needed. Therefore we have:
$v = \lambda f = (3 \text{ m})(2000 \text{ Hz}) = 6000 \text{ m/s}$

Interference

A mechanical wave is a disturbance that passes through a medium. As a harmonic disturbance moves, the particles of the medium react by oscillating about their equilibrium position with simple harmonic motion. **Interference** is the name used to describe what happens if two or more waves pass through the same region. The resultant displacement is the algebraic sum of the displacements produced by the individual waves. This rule for the addition of two or more waves is known as the **principle of superposition**.

If two waves add to give a resultant wave whose amplitude is less than that of either individual wave, the result is called **destructive interference**. If two waves add to produce a wave with an amplitude that is greater than either component, the result is called **constructive interference**.

If two waves combine such that their crests are aligned and their troughs are aligned, we say the waves are **in phase**. However, if the crests of one wave are aligned with the troughs of the other, we say the waves are **out of phase**. Other alignments are possible, and the phase difference between the two waves is generally expressed as an angle. Waves that are *out of phase* are said to have a phase angle of 180°, and waves that are *in phase* are said to have a phase angle of 0°. The sine and cosine functions are 90° out of phase.

A **standing wave** is an important example of interference that occurs when two waves of equal amplitudes and wavelengths combine as they travel *toward* each other. The resultant wave is known as a standing wave since the disturbance does not appear to move in either direction. A standing wave can be produced in a string by attaching one end firmly and shaking the other end. The wave produced by the shaking, travels down the string, reflects off the fixed end and travels back toward your hand. The pattern is the result of the wave traveling down the string from your hand combining with the reflected wave traveling in the opposite direction.

At certain locations on a standing wave, there will be no motion. These stationary points are called **nodes**. At a node, the two passing waves will *always* be 180° out of phase and will completely cancel out producing no motion. (There will be a node at the fixed end of the string because the reflected wave is always 180° out of phase with the incident wave.) **Antinodes** are places where the resultant motion is a maximum. Antinodes are located exactly half way between two adjacent nodes. Adjacent nodes (or adjacent antinodes) are separated by one *half* wavelength.

Resonance

As we have seen, wave motion and simple harmonic motion are intimately tied together. Many physical systems will vibrate at a certain frequency when disturbed. The frequency (or frequencies) at which a physical system vibrates when disturbed is called the **natural frequency** (or natural frequencies) of the system. When struck, a mass on a spring will vibrate at its natural frequency. When given a push, a child on a swing will swing back and forth (oscillate) with the natural frequency of the swing.

If a physical system is vibrated at or near one of its natural frequencies, the system will **resonate**. That is, it will oscillate, and the amplitude of the oscillations may become very large, even when the system is vibrated rather mildly. **Resonance** occurs when small amounts of energy are added to the system *in phase* with the motion of the system. The energy can only be added in phase with the motion if the source of the energy is vibrating near the natural frequency of the system. The added energy always *increases* the motion, leaving only friction to remove energy from the system.

In some cases of resonance, the amplitude of the motion can build to such a large value that the system is destroyed or badly damaged. This can happen to some buildings in an earthquake. As the seismic wave vibrates the building, if the frequency of the wave matches one of the natural frequencies of the building, the building may sway so violently that it is destroyed.

If the frequency of the seismic wave does not match a natural frequency of the building, at one instant the energy source (seismic wave) may be in phase with the motion of the building (causing an increase in the motion), but the next instant the energy source may be out of phase with the motion of the building (causing a decrease in the motion). The building will shake randomly, but will not resonate. That is, its amplitude will not build to a very large value.

Sound Waves

Sound waves are longitudinal waves or compression waves that we normally think of as traveling through air, but they can travel through any type of matter. Sound is produced by vibrating objects. As an object vibrates in air, the air is first compressed and then decompressed (rarefied) by the moving object. The sound wave is a disturbance that is composed of compressions (regions of higher pressure than atmospheric) and rarefactions (regions of lower than atmospheric pressure). These compressions and rarefactions travel outward from the source of the sound at a speed that depends on the temperature of the air and the molecular mass of the air. At a temperature of about 23°C, the speed of sound in air is about 345 m/s. The speed of sound in water is about 4 times faster (about 1440 m/s) and the speed of sound in most solids is even faster still (about 5000 m/s).

Sound is detected by the mechanism in the ear and our perception of the sound depends on the physical nature of the sound and our interpretation of the signal picked up by the ear. The **pitch** of a particular sound depends on the frequency of the wave. The **audible range** of frequencies that a *good* human ear is able to distinguish extends from about 20 Hz to about 20,000 Hz. Sound waves with frequencies below 20 Hz are called **infrasonic** waves, and sound waves with frequencies above 20,000 Hz are called **ultrasonic** waves. Ultrasound waves are used to examine internal tissue. The ultrasound waves are partially reflected at the boundaries of organs and other structures. The shape of these internal organs can be determined by studying the reflected ultrasound waves.

A wave carries energy, and the time rate at which the wave carries energy across a unit area is called the **intensity** of the wave. Intensity is measured in watts per square meter. Sound intensity *levels* are expressed using a logarithmic scale. The unit of this scale is the bel (B) or more commonly the decibel (abbreviated dB), which is a *tenth* of a bel. The **intensity level** (β) measured in *decibels* is given by:

$$\beta = 10 \log \frac{I}{I_o}$$
intensity level

Where I is the intensity of the sound and $I_o = 1 \times 10^{-12}$ W/m^2 is the threshold of hearing for an average person. If two sounds differ by 10 dB, then one is 10 times more intense than the other (10 log 10 = 10). If two sounds differ by 20 dB, one is 100 times more intense than the other (10 log 100 = 20). If two sounds differ by 30 dB, one is 1000 times more intense than the other (10 log 1000 = 30).

The intensity of sound that can be detected by the human ear depends on the *frequency* of the sound. Human ears do not respond as well to very low or very high frequencies, but respond best to frequencies between about 500 Hz and 5000 Hz. **Loudness** is the term used to describe how the intensity of a particular sound is *sensed* by a human being.

Sources of Sound

Strings such as those on a guitar or violin are fixed at both ends and are tightened to a given tension. When plucked, these strings will vibrate at certain natural frequencies and generate sound waves. The vibration of the string is the result of a standing wave being set up in the string. In the simplest situations, every point on the string vibrates at the same frequency.

There are many frequencies at which the string can vibrate, but the *lowest* frequency occurs when the string vibrates as one loop, the two fixed ends being nodes with an antinode at the center. The frequency of this vibration is called the **first harmonic** because the series of natural frequencies of a string of fixed length forms a harmonic series. Each successive frequency is an *integer* multiple of the first harmonic. The **second harmonic** has twice the frequency of the first harmonic, and it occurs when the string vibrates as two loops with a node at the center. The **third harmonic** has three times the frequency of the first harmonic, and it occurs when the string vibrates as three loops. At some point the frequency of the higher harmonics exceeds the frequency that can be heard by the human ear.

Generally, when the string of a guitar is plucked, the string vibrates in a complicated way that produces a *mixture* of harmonics. The strongest harmonic gives the pitch to the sound, but the **quality** or **timbre** of the sound is determined by the relative amplitudes of the various harmonics. You can notice a difference in the quality of the sound when a guitar string is plucked at different locations.

A plucked guitar string is *not* an example of resonance since energy is not fed into the system at a frequency near the frequency at which the string is oscillating. However, a violin bow drawn across a string is an example of resonance. In this case, the resin on the bow causes the bow to grab and release over and over again as the bow is drawn across the string. The bow gives the violin string many tiny plucks *in phase* with the motion of the string. Thus, energy is fed into the system at the natural frequency of the system and the string resonates.

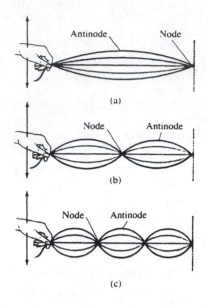

Figure 6.1 Standing waves in a string fixed at both ends. The first three harmonics are shown.

Sound waves will produce standing waves in long tubes. As a sound wave travels down a tube, a portion of the wave will be reflected at the other end. Even if the tube is open at both ends, there will be some reflection at the boundary and a standing sound wave will be produced in the tube. For a tube open at *both* ends, the tube will resonate or "sing" if the air near the ends of the tube vibrates with a very large amplitude. This requirement means that displacement antinodes must be located near the ends of the tube. Antinodes will be located near the ends of a tube open at both ends if the length of the tube (L) is some integer multiple of one half the wavelength of the sound wave traveling back and forth within the tube. In equation form this condition is:

$$L = \frac{n}{2}\lambda_n$$

Where n is an integer and λ_n is the wavelength of the sound wave. There are many wavelengths (and corresponding frequencies) that will cause the tube to resonate. These frequencies also happen to form a harmonic series. A tube can be made to resonate by holding a tuning fork (of the right frequency) near one end of the tube, or by blowing across the end of the tube.

Standing waves in strings or standing sound waves in long pipes are examples of interference produced when two waves of the same frequency travel in opposite directions. If two sound waves are produced by sources that are fairly close together, but their frequencies differ by only a few hertz, another type of interference known as **beats** will be produced. As the two sound waves travel out to an observer, they will be in phase at one instant, but a little later the waves will be out of phase, and still later they will be in phase again. The

82

observer will hear a loud sound when the waves are in phase, then little or no sound when the waves are out of phase. The observer will hear these loud and soft sounds in a regular pattern that we call **beats**. It can be shown that the frequency of the beats is just the difference in the frequencies of the two sound waves.

Example 6.3

A tube closed at one end will resonate when its length (L) and the wavelength of the sound wave are related by the expression: $L = n\lambda_n/4$
Where $n = 1,3,5,...$ For a certain tube closed at one end, consecutive resonances occur for wavelengths of 3 m and 2.4 m. Find the length of the tube.

Perhaps the best approach to a question like this one is to try several different n values until you get the same length for the two different wavelengths. First, a wave with a wavelength of $\lambda = 3$ m will resonate for the following lengths (L values):
(2)(3 m)/4=1.5 m; (3)(3 m)/4=2.25 m; (4)(3 m)/4=3 m
For $\lambda = 2.4$ m some of the lengths (L values) that will resonate are:
(3)(2.4 m)/4=1.8 m; (4)(2.4 m)/4=2.4 m; (5)(2.4 m)/4=3 m
The answer is 3 meters.

Doppler Effect

If the source of a sound wave is moving toward an observer, or the observer is moving toward the source, the frequency that the observer hears will be *higher* than the frequency emitted by the source. If the source is moving away from the observer, or the observer is moving away from the source, a *lower* frequency will be heard by the observer. This observed change in pitch due to the motion of the source or observer is called a **Doppler effect**. If a sound wave is *bounced off* a moving object, a Doppler shift in the frequency will also occur. This effect is used to measure the speed of blood flow in the human body. If the relative speed of the source and observer (v) is small compared to the speed of sound (v_s), the change in frequency (Δf) and change in wavelength ($\Delta\lambda$) are given *approximately* by:

$$\frac{\Delta f}{f} = \frac{\Delta\lambda}{\lambda} = \frac{v}{v_s} \qquad \text{(Doppler formula for } v<<v_s\text{)}$$

Where f and λ are the frequency and wavelength respectively that the *source* emits. If the source and observer are *separating*, the observed wavelength will be *longer* than the unshifted wavelength and the observed frequency will be *lower*. If the source and observer are *approaching*, the observed wavelength will be *longer* and the observed frequency will be *higher*. Doppler shifts also occur with light and other electromagnetic waves.

Example 6.4

A Doppler probe is used to monitor the flow of blood in a blood vessel. The probe uses ultrasound with a wavelength of 5×10^{-3} m and the speed of sound in human tissue is 1.5×10^3 m/s. If the blood in the vessel is moving toward the probe, the frequency of the reflected ultrasound wave is:

A. 3.0000×10^5 Hz

B. 3.0002×10^{-3} Hz

C. 2.9998×10^{-3} Hz

D. 7.5 Hz

Since the fluid is moving *toward* the source, the Doppler shift will cause an *increase* in the frequency. The frequency of the *unshifted* ultrasound wave is: $f = v/\lambda = (1.5 \times 10^3$ m/s$)/(5 \times 10^{-3}$ m$) = 3.0 \times 10^5$ Hz Therefore, the shifted frequency must be *greater* than this value. The only answer that fits this criterion is B.

When waves are bounced off a moving object and are detected by a receiver near or connected to the original source of the waves, there is a *double* Doppler shift. Doppler probes and the radar guns used to catch speeders operate this way. They contain *both* the source of the waves and the receiver of the reflected waves. The emitted wave is shifted since the source of the waves and observer are moving relative to each other. In this case the object off of which the wave is going to reflect can be considered as the observer. However, the reflected wave is shifted again. The reflected wave can be thought of as being emitted by a new source and the detector is the final observer.

Questions and Problems

1. Two mechanical waves with the same frequency pass through a certain region in space. One of the waves has an amplitude of 3 units and the other has an amplitude of 4 units. What are the possible values for the amplitude of the wave that results when the two waves combine?

2. A sound is 20 dB louder than a whisper. How many times greater is its intensity than a whisper?

3. If the length of a simple pendulum is doubled, by what percent does the period change?

4. The period of a mass on a spring is given by: $T = 2\pi\sqrt{\dfrac{m}{k}}$ A mass of 1 kg will produce a certain vibration frequency f_1. If the one kg mass is replaced by a mass of 4 kg, how is the new frequency f_4 related to f_1?
   A. $f_4 = 4f_1$
   B. $f_4 = 2f_1$
   C. $f_1 = 4f_4$
   D. $f_1 = 2f_4$

5. A flag pole is struck and the knob at the top of the pole begins to oscillate. Which of the following statements best explains why the motion might be simple harmonic?
   A. All oscillatory motion is simple harmonic.
   B. If the amplitude is small the motion will be simple harmonic.
   C. If the force is small enough the motion will be simple harmonic.
   D. Since it is not a simple pendulum it cannot be simple harmonic.

6. When the sound from a 440 Hz tuning fork is combined with a sound of unknown frequency, a beat frequency of 5 Hz is heard. What are the possible frequencies of the unknown sound? The frequency of the unknown sound is:
   A. 445 Hz or 435 Hz
   B. only 445 Hz
   C. 430 Hz
   D. 450 Hz

7. A tsunami (sometimes called a tidal wave) is a very long wavelength water wave. The wavelength of a certain tsunami is 5×10^5 m and it travels at a speed of 180 m/s in the open ocean. What is the time between incoming waves?

8. A pendulum bob of mass m is pulled to the side until the pendulum's string (whose length is L) is horizontal. After the bob is released, the speed of the bob at the lowest point in its swing is:

A. $2\pi\sqrt{\dfrac{L}{g}}$

B. $\sqrt{2gL}$

C. $\sqrt{gL}$

D. mgL

9. A tube open at both ends resonates with a fundamental frequency of 440 Hz and a second harmonic frequency of 880 Hz. How does the speed of the standing wave in the tube when it is resonating at the fundamental frequency compare with the speed of the wave when it is resonating at the second harmonic?

A. The higher frequency is produced by a faster moving wave.

B. The lower frequency is produced by a slower moving wave.

C. The wave is not moving since it is a standing wave.

D. The wave is moving at the same speed for all resonances.

Answers to Questions and Problems

1. If the waves are in phase the amplitudes add and produce an amplitude of $(3 + 4) = 7$ units, but if the waves are $180°$ out of phase the waves will partly cancel and produce an amplitude of $(4 - 3) = 1$ unit.

2. $20 \text{ dB} = 10 \log \dfrac{I_2}{I_1}$ so $\log \dfrac{I_2}{I_1} = 2$ and the ratio of the intensities is: $10^2 = 100$. The sound is 100 times the intensity of a whisper.

3. Since the period is proportional to the square root of the length, if the new length is made 2 times the old length, the new period will equal $\sqrt{2} = 1.4$ times the old period. The new period will be 40% longer than the old period.

4. You should realize that the period will *increase* (frequency will *decrease*) when a larger mass is added. This realization eliminates A and B. The one kg mass (call it *m*) is increased to $4m$. Since the period is proportional to the square root of the mass, the period changes from T_1 to $\sqrt{4}\,T_1 = 2\,T_1$. $(T_4 = 2T_1)$ The period is *doubled*. The frequency is inversely proportional to the period, therefore, the frequency is *halved*. f_4 is one half f_1 or $(f_1 = 2f_4)$. The answer is D.

5. Any *restoring* force will produce simple harmonic motion *if* the amplitude of the motion is small. The answer is B.

6. The beat frequency is equal to the *difference* of the two frequencies. Therefore, the unknown frequency could be 445 Hz = (440 Hz + 5 Hz) or it could be 435 Hz = (440 Hz - 5 Hz). The answer is A.

7. The frequency of the tsunami is given by:
$f = v / \lambda = (200 \text{ m/s})/(5 \times 10^5 \text{ m}) = 4 \times 10^{-4} \text{ Hz}$ The period (time between incoming wave crests) is the inverse of the frequency so:
period $= 1/(4 \times 10^{-4} \text{ Hz}) = 2500$ seconds.

8. Mechanical energy is conserved. If we let the potential energy at the lowest point be zero, the initial potential energy will then be mgL. The kinetic energy at the lowest point $(\tfrac{1}{2}mv^2)$ must equal the initial potential energy. Therefore we have: $mgL = \tfrac{1}{2}mv^2$
And: $v = \sqrt{2gL}$ The answer is B. Only B and C have the right units.

9. The wave bouncing back and forth in the tube is a *sound* wave and travels at the speed of sound (about 345 m/s at room temperature). The speed of sound does *not* depend on the frequency of the sound. The answer is D.

7 Electricity and Magnetism

7.1 The Electric Force

Atoms are composed of three different kinds of particles: neutrons, protons, and electrons. Electrons and protons have a property that we call **electric charge**. There are two kinds of electric charge, one called positive and one called negative. Objects that posses the same kind of charge are observed to repel each other, but objects with different charges will attract each other. The unit of electric charge is the coulomb (abbreviated C). **Point charges** are charges that are very small compared to the distance separating them. The *magnitude* of the force of attraction or repulsion on two *point* charges is given by **Coulomb's law**:

$$F = k\frac{Q_1 Q_2}{r^2}$$ *Coulomb's law*

Were Q_1 is the magnitude of the charge on one of the objects, Q_2 is the magnitude of the charge on the other, r is the distance between the objects, and k is a proportionality constant whose value is about:

$$k = 9 \times 10^9 \text{ N·m}^2/\text{C}^2$$

Except for the sign, the charge on an electron is numerically equal to the charge held by a proton. This fundamental unit of charge is called the **elementary charge** and is represented by the symbol e:

$$e = 1.6 \times 10^{-19} \text{ C}$$ *elementary charge*

Since the negative electric charge on an electron exactly cancels the positive charge on a proton, normal matter with equal numbers of electrons and protons is electrically neutral. Protons are almost 2000 times more massive than electrons, so it is much easier to move an electron than a proton. Therefore, when electric charges move, electrons are generally the charges that move. A positive charge is generally a *deficiency* of electrons and a negative charge is an *excess* of electrons.

Certain objects easily build up electric charges. For example, as a comb runs through a person's hair, electrons are rubbed off the hair by the comb and the comb acquires a negative charge (an excess of electrons). The hair acquires a positive charge since it now has a deficiency of electrons.

Although it may be easy for a comb to collect electrons, the electrons do not move about the comb easily and the comb tends to remain charged. An **insulator** is a material on which electric charges find it difficult to move about. Other materials, called **conductors**, allow charges to move around freely within the material. Insulators are materials in which the electrons are tightly bound to the nuclei, but conductors have many electrons that are only loosely bound to nuclei and are free to move about in the material. Metals are good conductors, while glass, rubber, wood, and plastic are insulators. The "metallic look" of metals is produced by the free electrons that efficiently reflect most of the incident light.

Example 7.1

Two charged particles are separated by a distance of 2 meters. If the distance between the two particles is doubled and the charges on the particles are doubled, how does the force between the two particles change?

Before the doublings, the force on each charge was given by: $F = k\dfrac{Q_1 Q_2}{r^2}$

To calculate the force after the changes have occurred, we must replace Q_1 with $2Q_1$, Q_2 with $2Q_2$, and r with $2r$. The new force is:

$$F_{new} = k\dfrac{(2Q_1)(2Q_2)}{(2r)^2} = k\dfrac{Q_1 Q_2}{r^2}$$

But this force is the same as the original force, therefore, the force did not change.

7.2 Electric Fields

If one electric charge is brought near another electric charge, it feels a force, but how does one charge know the other is near? To help us visualize what is happening, we imagine that an electric charge sets up a **field** in the space around itself that is somehow detected by other objects that possess an electric charge.

In a similar fashion, we imagine that objects with mass, like the Earth or a rock set up a different kind of field (some kind of field related to gravity) so other objects with mass will know that it is near. This gravity field could be a *force* field, and it could be investigated by using a "test mass" like a one kilogram brick. The brick could be placed at various locations around Earth and the force on the brick could be recorded. In this manner, the force field around Earth could be mapped.

The problem with a *force* field is that its value depends on the mass of the "test particle" (the brick). A two kilogram brick would have twice the force as a one kilogram brick. Therefore, we generally describe a gravitational field as the force divided by the mass of the test particle. The units of this gravitational field are force/mass or acceleration. Near Earth's surface, its gravitational field has a value of about 9.8 N/kg or 9.8 m/s^2. To find the force Earth exerts on a particle, we multiply the gravitational field strength by the mass of the particle, a quantity we call the weight of the particle.

In a similar way, the strength of a *force* field surrounding an electric charge will depend on the electric charge of the test particle. To avoid this dependence on the charge of the test particle, we define an electric field to be the force divided by the electric charge of the test particle. If you want to find the force on an electric charge placed in an electric field, you multiply the electric field strength by the charge on the particle.

To summarize, the **electric field strength** at a point in space is defined to be:

$$\mathbf{E} = \mathbf{F}/q \qquad\qquad\qquad\qquad \textit{electric field}$$

Where q is the electric charge of the test charge (taken to be a *positive* test charge) and $\mathbf{F}$ is the force exerted on the test charge by the electric field. Electric fields are measured in units of newtons/coulomb (N/C). The electric field is a *vector* field, which means that we imagine that each point in space has a vector associated with it that gives the strength and direction of the electric field. The direction of the field is the direction of the force vector on the *positive* test charge.

The electric field is often represented in a visual way by drawing lines in the space around an electric charge that are *tangent* to the electric field vectors. These **electric field lines** are "down hill lines" in the sense that they indicate the direction in which a *positive* charge will *begin* to move if it is placed at some point in space. They do *not* show the actual path on which a charged particle will move, only the way it will *start* to move if placed *at rest* at some point.

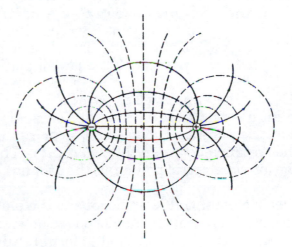

The force on any electric charge q placed in an electric field $\mathbf{E}$ can be calculated using the expression:

$$\mathbf{F} = q\mathbf{E}$$

Figure 7.1 Two particles with opposite charges (called an electric dipole) showing the electric field lines (solid) and equipotential lines (dashed).

Where $\mathbf{E}$ is the electric field strength at that point in space. If the electric charge is a negative, the direction of the force exerted by the field will be opposite to the direction of the electric field.

In static situations (where no charges are moving) the electric field inside a conductor will be zero. This is true because charges are free to move in a conductor, so they will quickly move to set up a charge distribution that will cancel any field that might have been present initially in the conductor. Because charges are free to move in conductors, any excess electric charges on a conductor must be on the *surface* of the conductor in the static situation. Any net charge placed inside the conductor would produce an electric field inside the conductor and cause charges to move until the electric field inside becomes zero, and the net charge inside the conductor becomes zero.

7.3 Electric Potential Difference

Since an electric field exerts a force on an electrically charged particle, if the particle moves from one point to another (say from point a to point b), the field will do work on the electric charge. We can define an electric potential energy difference for charged particles in a manner similar to the way we did with gravity, but a more useful quantity is the electric potential energy difference *per unit charge* which is called the **electric potential difference**, often shortened to the electric potential. (Notice it does not contain the word *energy*.)

Electric potentials and potential energies are always *differences*. When we talk about the "potential energy" of a particle, it is always the potential energy *relative* to some zero point. For gravity, we took the potential energy to be zero at the surface of the Earth. For electric potentials, we generally define the zero point to be the point with the lowest potential value, and we call it the **ground**.

Electric field lines will be *parallel* in a region where the electric field is *constant*. If an electric field is constant or nearly so, the electric potential difference (electric potential) between two points is just the electric field strength times the distance between the two points. The electric field points to the point with the *lowest* electric potential (the ground).

Electric potential is analogous to the elevations on a topographic map, and the electric field lines are analogous to down hill lines. Points with the same electric potential can be connected to form **equipotential lines**, which are analogous to the contour lines on a topographic map. Where they intersect, equipotential lines are *perpendicular* to electric field lines (Fig. 7.1).

Electric potential difference is measured in units of joules/coulomb or volts (V). One volt is equal to one joule per coulomb. In a *uniform* electric field, the electric potential difference between two points separated by a distance d is given by:

$$V = Ed \qquad \text{(for uniform electric field)}$$

If a particle at rest with a charge of q is accelerated by an electric potential difference of V, the particle will gain kinetic energy. Its gain in kinetic energy (K) is given by:

$$K = qV$$

Where qV is the particle's loss of potential *energy* as it is accelerated by the electric potential difference V. Electrons in cathode ray tubes, X-ray machines, and many other instruments are accelerated through high voltages (electric potential differences) to create beams of high energy electrons. Often we measure the kinetic energy of these electrons using a unit of energy called the electron volt. One electron volt is the kinetic energy gained by an electron after it has been accelerated through an electric potential difference of one volt. Since the charge on an electron is 1.6×10^{-19} C, the electron volt (abbreviated eV) and the joule are related as follows:

$$1 \text{ eV} = (1.6 \times 10^{-19} \text{ C})(1 \text{ V}) = 1.6 \times 10^{-19} \text{ J} \qquad \textit{electron volt}$$

Common devices used to maintain electric potential differences are batteries and electrical generators. Electric companies use generators to maintain the electric potential difference between the two holes in the electrical outlets in our homes.

7.4 Capacitance

A **capacitor** (sometimes called a condenser) is a device used to store electrical charge. A typical capacitor is composed of two conductive plates that are placed near each other but are not touching. When hooked up to a source of electric potential difference such as a battery, charge is transferred from one plate to the other. The charge transferred Q is proportional to the voltage difference V across the plates, and the constant of proportionality is called the capacitance C. In equation form we write:

$$Q = CV \qquad \textit{capacitance}$$

Capacitance is measured in units called farads (abbreviated F). One farad is equivalent to one coulomb per volt. The capacitance of a device is proportional to the area of the plates and is inversely proportional to the separation distance of the plates.

Capacitance also depends on the type of material that is placed between the plates. Materials called dielectrics are often placed between the plates of capacitors to *increase* the capacitance. If a dielectric material is inserted between the plates of a capacitor with a charge Q on its plates, the dielectric will not change the charge on the plates, but the material *decreases* the voltage between the plates which *increases* the capacitance of the device.

Strong *uniform* electric fields can be created in the spaces between the plates of charged capacitors. The electric field of a capacitor can be used to accelerate the electrons in cathode ray tubes and X-ray machines. Such a capacitor is called an electron gun. The negative plate is called the **cathode** and the positive plate is called the **anode**. In the early days before people knew what they were (and even after they knew), particles emitted by the cathode were called cathode rays.

7.5 Electric Currents and Resistance

A battery is a common device that will maintain an electric potential difference between two points called the terminals of the battery. The electric potential at one of the terminals (called the positive terminal) is higher than the electric potential at the other terminal (the negative terminal). It is usually understood that the electric potential of the negative terminal is zero or ground, and we often just give the electric potential of the positive terminal. For example, we talk about 12 volt batteries, which are batteries with an electric potential *difference* of 12 volts between the two terminals of the battery.

The difference in electric potential is called the **electromotive force** (emf) of the battery. (It is called an electromotive force even though it is *not* a force but a voltage.) If a conductor is connected between the terminals of the battery, an electric field will be set up in the conductor causing electric charges to move through the conductor. The electric charges (usually electrons) move on a closed path through the conductor and the battery. This continuous conducting path is called a **circuit**.

A flow of electric charge is called a **current**. The amount of current flowing is defined to be the amount of charge that passes through a cross-sectional area of the conductor per unit time. Electric current is measured in coulombs per second. One coulomb per second is called an **ampere** (abbreviated A). Since electric current is easier to measure than electric charge, the ampere is a fundamental unit and the coulomb is a derived unit. The coulomb is defined to be the amount of charge that flows through a given cross-sectional area in one second when a current of one ampere is flowing.

In most conductors it is the negatively charged electrons that move since they are much less massive than the positive nuclei. However, we define the direction of an electrical current to be the direction positive charges *would* flow *if* they were moving. This current is the **conventional current** in a circuit.

When a potential difference is maintained between the ends of a conducting device, the amount of current flowing depends on the nature of the device. For a given voltage, some devices do not allow as much current to flow as others. There is a certain amount of "resistance" to the flow of current in all devices. The electrical **resistance** R of a device is defined by the following expression:

Where I is the flow of electrical current through the device when a potential difference of V is maintained across the device. The unit of electrical resistance is the **ohm**, and one ohm is equal to one volt per ampere. The abbreviation for the ohm is the capital Greek letter omega (Ω). Any device with electrical resistance is often simply called a resistor. The definition of resistance is generally written as:

$V = IR$

Georg Ohm discovered that for many devices, the resistance of the device is *approximately* constant. This statement, which is only approximately true for many conductors, is called Ohm's law. Fortunately, Ohm's law is not even approximately true for many of the electronic components that are essential in producing modern computers and electronic equipment. Even though Ohm's law is not generally true, the relationship between potential difference, current, and resistance is true since it is the *definition* of resistance. Many people *erroneously* refer to $V = IR$ as Ohm's law.

A wire will have a certain resistance to the flow of electricity. The resistance of a wire depends on three things: it is directly proportional to the length of the wire (longer wires have a greater resistance); it is inversely proportional to the cross sectional area of the wire (fatter wires have less resistance); and it will depend on the type of material the wire is made of. We can write an equation for the resistance (R) of a wire of length L and cross-sectional area A as:

$R = \rho L/A$

Where ρ is a quantity called the **resistivity** of the material from which the wire was made. (Even though we use the same symbol as we did for density, the resistivity has nothing to do with the density of the material.) The resistivity is an intrinsic property of the material and does *not* depend on the *geometry* of the resistor. Copper has a relatively *low* resistivity which means it is a good material for making the wires that current flows through on its way to electrical devices. Aluminum is not as good, but it is a much cheaper material so high power lines are generally made of aluminum. Its higher resistivity can be compensated for by making the aluminum wires a little thicker.

7.6 Electric Power

When an electric charge Q flows through a device with a certain resistance, the charge looses electric potential energy. For a pure resistor, this loss shows up as increased internal energy (the resistor gets warmer). The voltage across the resistor V is the *change* in electric potential energy per unit charge, so the loss in potential energy of charge Q is just QV. The time rate at which potential energy is lost is just the potential energy divided by the time. But the charge divided by the time is the current I. So, the **power** used by a resistive device is:

$$P = QV/t = IV$$

Since the voltage across the resistive device is IR, the power used by a resistive device can also be written as:

$$P = IV = I^2R = V^2/R \qquad\qquad \textit{electric power}$$

7.7 Alternating Current

Current supplied by electrical companies is alternating current (abbreviated AC). The voltage maintained by the company changes direction 120 times each second and causes the electrons in a conductor to execute simple harmonic motion with a period of 60 Hz. Although the electrons in an AC circuit oscillate about their equilibrium position rather than move through the circuit like a stream of water, the expressions for resistance and electrical power still apply. The appropriate values that must be used for the voltage and current are *averages* called **root-mean-square** (rms) values. The rms voltage of household electrical outlets varies with location, but it is approximately 120 volts. The power expression gives the *average* power used by the AC device.

Example 7.3

How much power can be drawn from an electrical outlet if the circuit breaker is rated at 15 amps?

The 15 amp current is an rms value. Since a typical value for the rms voltage is 120 volts, the power that can be drawn is:
$P = IV = (15\ A)(120\ V) = 1800\ W$. Now you know why it is not a good idea to run more than one hair dryer from the same electrical outlet.

7.8 Parallel and Series Circuits

If two or more electrical devices are connected end-to-end so the *same current* passes through each device, the devices are said to be connected in **series**. Devices connected in series have the same current passing through them, but the potential drop across each device will be different depending on the resistance of the device. The potential drop across a given *element* in the circuit is the product of the current through the circuit and the resistance of the *element* ($V = IR$).

Generally, series connections are found in limited applications. Some Christmas tree lights are connected in series so if one burns out the whole string goes off. If you wish to measure the current through a device, a current meter must be connected in *series* with the device so the same current runs through the meter and the device. Hopefully, the resistance of the meter is small compared to the resistance of the device, otherwise its presence will reduce the current significantly.

Most devices, however, are designed to run at or near some specific voltage and they must be connected to a power supply (a voltage source) in parallel. **Parallel** connections are made so the *voltage* across each device is the same. This is accomplished by connecting the high potential sides of all the devices together (and to the high voltage terminal of the power supply), and connecting the low potential sides of all the devices together (and to the low voltage terminal or *ground* of the power supply).

Devices connected in parallel will have the same potential difference across their leads, but the *current* through each will depend on the resistance of the particular device. Since the voltage is the same, the current through a given device is *inversely* proportional to its resistance ($V = IR$). A device with a *smaller* resistance will draw *more* current than a device with a higher resistance. When you plug several devices into electrical outlets you are connecting them in parallel since each device is connected to the same voltage (approximately 120 volts in the case of a normal electrical outlet).

Example 7.5

A 1 Ω and a 2 Ω resistor are connected in parallel across a power supply. If the resistors draw a current of 0.5 A from the power supply, what is the voltage across the 1 Ω resistor? Hint: The MCAT is designed to test your reasoning ability, not how many formulas you have memorized.
A. 0.18 V
B. 0.33 V
C. 0.50 V
D. 1.50 V

Do not jump for a formula, use your reasoning ability. A larger current will flow through the 1 Ω resistor. If the entire 0.5 A went through the 1 Ω resistor, the voltage across it would be: $V = IR = (0.5 \text{ A})(1 \text{ Ω}) = 0.5 \text{ V}$. Therefore, C and D must be incorrect since they are too large. Also, the current through the 1 Ω resistor must be larger than *half* the total current (more than half of 0.5 A or more than 0.25 A). 0.25 A through the 1 Ω resistor would give a voltage of 0.25 V so the only answer that is larger than 0.25 V and less than 0.50 V is B.

Another approach would be to let I be the current through the 1 Ω resistor and then the current through the 2 Ω resistor would be (0.5 A - I). Since the voltage is the same across each resistor, we equate the IR drops across each:

$I(1\ \Omega) = (0.5\ A - I)(2\ \Omega)$

$I = 1\ A - 2I$

$I = 0.33\ A$

7.9 The Internal Resistance of a Battery

Batteries and power supplies are sources of emf. Batteries maintain this emf by chemical means. A battery has an **internal resistance** so when current is drawn from the battery, the potential difference across the terminals of the battery (called the **terminal voltage**) *drops* by an amount Ir, where I is the current drawn and r is the internal resistance of the battery. This relationship can be written as:

$$V = \mathcal{E} - Ir \qquad\qquad\qquad \textit{terminal voltage}$$

Where V is the terminal voltage and $\mathcal{E}$ is the emf of the battery (the voltage when *no* current is flowing). Dry cell batteries have a *higher* internal resistance than wet cells. This higher internal resistance explains why you cannot start your car with a 12 volt dry cell. When the car starter tries to draw a large current, the terminal voltage drops to near zero.

7.10 Magnetism

Magnets will attract pieces of iron and will interact with other magnets. Just as we envisioned an electric field around a charged particle and a gravitational field around an object with mass, we envision that the space surrounding a magnet is filled with a **magnetic field** that interacts with certain "magnetic" objects.

Magnets *always* have two poles, a north pole and a south pole. A magnetic field can be mapped by using a small magnet as a probe. If the small probe magnet is suspended so it can rotate freely, it will align itself with any magnetic field in which it is placed. The north pole of one magnet will be attracted to the south pole of another, and like poles will repel each other. We envision the **magnetic field lines** coming out of the north pole of a magnet and entering the south pole.

Earth's magnetic field can be mapped by using a compass whose needle is completely free to move in all directions. The magnetic field of Earth points out of Antarctica (which is where the *north* pole of the Earth's magnetic field is located) and the field points into the Arctic (which is the location of the *south* pole of the Earth's magnetic field). The North Magnetic Pole of the Earth is really the pole the north end of a magnet "seeks", or the north seeking pole. Earth's magnetic field is parallel to the ground at the equator.

Electricity and magnetism are two aspects of a single interaction in nature called the electromagnetic interaction. Magnetic fields are produced by moving electric charges, and they disappear if one travels along with the electric charge.

An electric current moving along a wire will cause a magnetic field to be created around the wire (Fig. 7.2). The magnetic field lines are circles centered on the wire. The direction of the magnetic field lines can be determined by the **right-hand-rule**. If you align the thumb of your right hand with the direction of the conventional current (the direction positive charges would flow), your fingers will curl in the direction of the magnetic field. The strength of the magnetic field is proportional to the current in the wire and gets weaker as you move away from the wire. (For a long straight wire it is inversely proportional to the distance from the wire.)

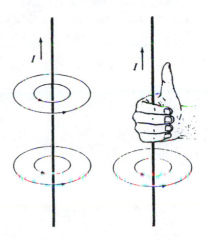

Figure 7.2 Magnetic field around a long straight current carrying wire.

A loop of wire with current flowing around it will produce a magnetic field that looks similar to the magnetic field surrounding a short bar magnet (Fig. 7.3). The direction of the magnetic field lines can be found from the right-hand-rule. A long wire can be wound in many loops to produce a stronger magnet (this increases the current by increasing the number of loops).

The magnetic field of a permanent magnet is produced by the atoms that make-up the material. Certain atoms like iron and nickel have very small magnetic fields, and if many of the atoms line up in a solid, the solid will have a strong magnetic field. If a piece of iron is placed in a current loop, the strength of the magnetic field will be greatly increased since the iron atoms (which act like little magnets) will tend to line up with the magnetic field produced by the coil and reinforce the strength of the field.

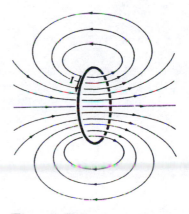

Figure 7.3 The magnetic field produced by current flowing around a circular wire.

A charged particle moving through a region of space where a magnetic field is present will experience a force due to the magnetic field if the velocity of the particle is *not* along a magnetic field line. The force on the particle will be *perpendicular* to *both* the magnetic field and the charged particle's velocity vector. It will experience a maximum force if it is moving in a direction perpendicular to the magnetic field. The force on the charge depends on the size of the charge q, the velocity of the charge v, and the strength of the magnetic field B. The expression for the *maximum* force is:

$$F = qvB \quad \textit{(maximum force on a moving charge in a magnetic field)}$$

Magnetic fields are measured in units of newtons/(coulombs)(meter/sec) or newtons/(ampere·meter). One N/(A·m) is called a tesla (abbreviated T).

Questions and Problems

1. Three charged particles (A, B, and C) are located at the vertices of an equilateral triangle as shown in the figure. The magnitudes of the three charges are equal and their sign is indicated on the figure. The direction of the force on the top charge is:

    A. Up
    B. Down
    C. Left
    D. Right

    A $\oplus$

    B $\oplus$ C $\ominus$

2. A capacitor can create a uniform electric field of 2×10^6 N/C in the region between its plates. If an electron is accelerated by this field, find the kinetic energy gained by the electron if the plates are 1 mm apart.

3. The distance between two electrons in an electron beam is doubled. By what factor is the force between the two electrons changed?

4. In a region of space that contains a magnetic field, a magnet will align itself with the magnetic field. This alignment occurs because the magnetic field applies a torque on the magnet. The torque on the magnet is given by $\tau = \mu B \sin\theta$ where B is the magnetic field strength, μ is the magnetic dipole moment, and θ is the angle between the magnetic field and the magnetic dipole moment. If a magnetic with a dipole moment of 2.5×10^3 N·m^3/Wb is placed at an angle of $90°$ to a magnetic field with a strength of 0.02 Wb/m^2, find the torque on the magnet.

5. What is the resistance of a 60 W light bulb?

6. How much current would be required to run four 5 kilowatt heaters if a single 500 volt power supply is to be used?

7. An electron has a mass of 9.1×10^{-31} kg and a charge of 1.6×10^{-19} C. In a fluorescent lamp, electrons encounter an electric field of 6000 V/m. Find the approximate acceleration of an electron in one of these lamps.

    A. 1×10^{15} m/s^2
    B. 5×10^{15} m/s^2
    C. 7×10^{15} m/s^2
    D. 9×10^{15} m/s^2

8. In an electric field, positive ions will flow in the direction of the field and electrons will flow in the direction opposite to the direction of the field. If an electron and a positive ion have electric charges of equal magnitude, why will the electrons travel faster than the ions?

101

9. Electrons (with a charge of 1.6×10^{-19} C) are fired by the electron gun in an electronic instrument. The gun runs at a voltage of 10,000 V and sends a current of 0.032 A to the screen of the instrument. How many electrons are emitted from the cathode of the gun each second?

10. A vertical antenna creates radio waves when an oscillating electric current is generated in the antenna. What is the direction of the magnetic field?
    A. Parallel to the ground and parallel to the antenna.
    B. Parallel to the ground and perpendicular to the antenna.
    C. Perpendicular to the ground and perpendicular to the antenna.
    D. Perpendicular to the ground and parallel to the antenna.

11. X-rays are produced when a high energy electron beam hits a metal anode. A certain X-ray tube operating at 10^5 V must maintain an electron current of 0.006 A in order to produce the desired X-ray intensity. How much power will the tube consume?

12. Blood contains ions that can be utilized by monitoring devices. If the pole of a magnet is pressed against an artery so the blood flow is perpendicular to the magnetic field lines, the ions will experience a force. The direction of the force is:
    A. Perpendicular to the velocity and parallel to the magnetic field.
    B. Parallel to the velocity and parallel to the magnetic field.
    C. Perpendicular to the velocity and perpendicular to the magnetic field.
    D. Parallel to the velocity and perpendicular to the magnetic field.

13. A battery is used to run a heating coil that is immersed in a cup of water. Which of the following best describes the energy transfers that take place?
    A. Electrical to heat to chemical
    B. Electrical to chemical to heat
    C. Chemical to electrical to heat
    D. Chemical to heat to electrical

14. A power supply is used to produce ultrasound waves that bounce off organs in the body. A transducer detects the waves and sends a signal to a display monitor. Which of the following best describes the energy transfers that take place in this series of steps?
    A. Electrical to mechanical to chemical
    B. Electrical to chemical to mechanical
    C. Electrical to heat to chemical
    D. Electrical to mechanical to electrical

15. A 6 V battery is connected to a 12 Ω and an 8 Ω resistor in parallel. How much current is flowing through the 8 Ω resistor?

16. Earth's magnetic field is very strong at the North Magnetic Pole, however, a compass does not work well in this region. Which of the following best explains why.
A. Earth's magnetic field near the pole is too strong for the compass.
B. Earth's magnetic field is parallel to the ground at the equator.
C. The vertical component of the field is too small.
D. The horizontal component of the field is too small.

17. The resistivity of copper is about 1.6×10^{-8} $\Omega \cdot m$ and the resistivity of aluminum is about 2.5×10^{-8} $\Omega \cdot m$. What must be the diameter of an aluminum wire if its resistance per unit length is to be the same as a copper wire with a diameter of 1 mm?
A. 0.64 mm
B. 0.80 mm
C. 1.25 mm
D. 1.56 mm

Answers to Questions and Problems

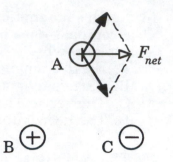

1. There are two forces on the top charge, a force due to B and a force due to C. The two forces are of equal magnitude since the charges and distances are the same. Charge A is repelled by charge B and it is attracted by charge C. The vector sum of these two equal forces points to the right.

2. The electric potential difference between the two plates of the capacitor is:
    $V = Ed = (2 \times 10^6 \text{ N/C})(1 \times 10^{-3} \text{ m}) = 2 \times 10^3 \text{ V}$ Therefore, an electron will gain 2×10^3 eV of kinetic energy or:
    $(2 \times 10^3 \text{ eV})(1.6 \times 10^{-19} \text{ J/eV}) = 3.2 \times 10^{-16} \text{ J}$ of kinetic energy.

3. The force is obviously decreased. Let the old force be $F = k\dfrac{Q_1 Q_2}{r^2}$. To find the new force, we must replace r by $2r$. Therefore, the new force is
    $F_{new} = k\dfrac{Q_1 Q_2}{(2r)^2} = \dfrac{1}{4}k\dfrac{Q_1 Q_2}{r^2} = \dfrac{1}{4}F$. The force decreased by a factor of 1/4.

4. Since the sine of 90° is 1, the torque is:
    $\tau = (2.5 \times 10^3 \text{ N}\cdot\text{m}^3/\text{Wb})(0.02 \text{ Wb/m}^2) = 50 \text{ N}\cdot\text{m}$. Wb is an abbreviation for webers, a unit of magnetic flux. $1 \text{ Wb/m}^2 = 1$ T. However, you did not need to know that to solve the problem.

5. Since a typical value for the voltage at which the light bulb operates is 120 V, we can use $P = IV$ to find the current flowing through the light bulb. $I = P/V = (60 \text{ W})/(120 \text{ V}) = 0.5 \text{ A}$. We then use $V = IR$ to find the resistance. $R = (120 \text{ V})/(0.5 \text{ A}) = 240 \ \Omega$

6. The total power required for the 4 heaters is $4 \times 5 \text{ kW} = 20 \text{ kW}$, and we use $P = IV$ to find the current required. $I = P/V = (20 \text{ kW})/(500 \text{ V}) = 40 \text{ A}$

7. The force on the electron can be found using $F = qE$ and the acceleration can be obtained by dividing the force by the mass ($F = ma$). Therefore:
    $a = qE/m = (1.6 \times 10^{-19} \text{ C})(6000 \text{ V/m})/(9.1 \times 10^{-31} \text{ kg}) \approx 1 \times 10^{15} \text{ m/s}^2$
    The answer is A.

8. The magnitude of the *force* on the electron and the *force* on the ion will be the *same*. But for a given force, the *acceleration* of a particle is inversely proportional to its mass, and ions are much more massive than electrons. Therefore, the electrons will have a greater acceleration and attain a faster speed than the positive ions.

9. Since the current is the charge divided by the time, the charge arriving at the screen in one second is just (0.032 A)(1 s) = 0.032 C. The number of electrons (n) needed to produce an electric charge of 0.032 C is:

$n = (0.032 \text{ C})/1.6 \times 10^{-19} \text{ C} = 2 \times 10^{17}$ The voltage is not needed.

10. The magnetic field is produced by the current in the antenna. Using the right-hand-rule, the field lines must be circles centered on the antenna. They must by parallel to the ground (the antenna is a vertical wire), and they must also be perpendicular to the antenna. The answer is B.

11. We use the equation $P = IV$ since we are given the voltage and the current.

$P = (0.006 \text{ A})(10^5 \text{ V}) = 6 \times 10^2 \text{ W}$

12. The magnetic force on a moving charge is perpendicular to *both* the magnetic field and the velocity vector of the charge. The answer is C.

13. The battery changes chemical energy to electrical energy, which is changed to heat (really internal energy) in the coil. Heat is the name we give the energy transferred from the coil to the water. The answer is C.

14. The power supply converts electrical energy to ultrasound waves that are mechanical energy. This mechanical energy is converted to electrical energy by the transducer. The answer is D. A transducer is a device that converts one form of energy to another. For example, a transducer can convert thermal energy under your tongue to electrical energy (an electric current) that can be measured to determine your temperature.

15. Since they are connected in parallel, the voltage across each resistor is 6 V. We use $V = IR$ to find the current through the resistor.

$I = V/R = (6 \text{ V})/(8 \text{ } \Omega) = 0.75 \text{ A}$

16. A compass needle is constrained to move in a horizontal plane, and it aligns itself with the horizontal component of the Earth's magnetic field. Earth's magnetic field is perpendicular to the ground at the North Magnetic Pole, so it has no horizontal component with which the compass needle can align. The answer is D.

17. The resistance is *directly* proportional to the resistivity and *inversely* proportional to the cross-sectional area, therefore, the aluminum wire will have to be *fatter*. The answer is either C or D. The cross-sectional area is proportional to the square of the diameter. Therefore, for the resistances to be equal, the resistivity divided by the square of the diameter (ρ/D^2) must be equal for both wires. If we let D be the diameter of the aluminum wire we have: $(1.6 \times 10^{-8} \text{ } \Omega\cdot\text{m})/(1 \text{ mm})^2 = (2.5 \times 10^{-8} \text{ } \Omega\cdot\text{m})/D^2$

$D^2 = (1 \text{ mm})^2(2.5 \times 10^{-8} \text{ } \Omega\cdot\text{m})/(1.6 \times 10^{-8} \text{ } \Omega\cdot\text{m}) = (1 \text{ mm})^2(25/16)$

$D = (1 \text{ mm})(5/4) = 1.25 \text{ mm}$ The answer is C.

8 Light and Optics

8.1 Electromagnetic Waves

As we saw in our study of electricity, the space surrounding a charged particle (like an electron) is filled with an electric field. The electric field lines can be visualized as strings or rubber bands extending out from the electron. If the electron were to wiggle or oscillate, the field lines would be shaken and waves would travel out along the field lines just as waves travel out along a string as you shake one end. These disturbances traveling out along the electric field lines are called **electromagnetic waves**. (The name comes from the fact that there is also a magnetic field disturbance created by the oscillating electric charge. *Moving* electric charges create magnetic fields.) Electromagnetic waves are sometimes called electromagnetic radiation.

The changing electric field is *parallel* to the direction of motion of the oscillating charge, and it is *perpendicular* to the magnetic field. (Fig. 8.1) Electromagnetic waves are transverse waves since both the electric and magnetic fields are perpendicular to the direction in which the wave is traveling. In a vacuum, all electromagnetic waves travel at the same speed. This speed, called the **speed of light** in a vacuum is usually represented by the symbol c.

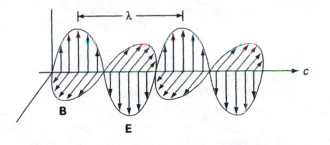

Figure 8.1 The electromagnetic wave moving to the right is produced by a charge oscillating up and down in the figure. The electric field vectors (**E**) and magnetic field vectors (**B**) are perpendicular to each other and to the direction in which the wave is moving.

$$c = 3 \times 10^8 \text{ m/s} \qquad\qquad \textit{speed of light}$$

As with any wave, the velocity of the wave is the product of the frequency (the number of wavelengths that pass by each second) and the wavelength (the length of one wavelength).

$$c = \lambda f$$

Electromagnetic (EM) waves are given different names depending on their wavelength. EM waves with the shortest wavelength (highest frequency) are called gamma rays, next comes X-rays, then ultraviolet light, visible light, infrared light, microwaves and radio waves. **Visible light** waves have wavelengths between about 400 nm (violet) and 750 nm (red).

Although electromagnetic waves are similar to waves on a string, they are localized disturbances that carry a specific amount of energy. This localized bundle of energy is called a **photon**. The photon has properties that are more easily explained by treating it as a particle. The energy of a photon is proportional to its frequency (it is inversely proportional to the wavelength of the photon). The energy of a photon is given by:

$$E = hf = hc / \lambda \qquad \qquad energy\ of\ a\ photon$$

Where h is a constant called **Planck's constant** and has a value of:

$$h = 6.6 \times 10^{-34} \text{ J·s} \qquad \qquad Planck's\ constant$$

Certain phenomenon are best explained by treating light as a particle while other effects are best explained by treating light as a wave. Because we use these two models to analyze light, we often say there is a **dual nature of light**.

The fact that ultraviolet photons have much more energy than visible light photons explains why ultraviolet light is so much more dangerous than visible light or infrared light. An ultraviolet photon has enough energy to rupture a small capillary but a visible photon does not. A sunburn results when many capillaries are ruptured and blood accumulates under the skin. Exposure to ultraviolet light will cause a sunburn, but you cannot get a sunburn from visible or infrared light. Just as a small caliber bullet does not have enough energy to pierce thick armor, a visible photon cannot rupture a capillary.

X-ray photons have even more energy than ultraviolet photons and they are able to penetrate deep into the human body. When absorbed or scattered, they can cause severe alterations in DNA molecules. These changes can cause the cell to become cancerous. Infrared or visible light photons do not have enough energy to produce this type of change, even on the skin.

An object that absorbs light will also emit light, but the light emitted by an object a room temperature is generally infrared light. An exception to this general rule is found in **fluorescent** material. These substances absorb ultraviolet photons and emit visible photons. Since ultraviolet photons are invisible to humans, fluorescent materials seem to be producing visible light at a low temperature, when in fact they are just converting the energy in ultraviolet photons into lower energy visible light photons.

Example 8.1

The mercury vapor in a fluorescent lamp emits ultraviolet photons with a wavelength of 200 nm, but the fluorescent material on the walls of the lamp converts many of these UV photons into 550 nm visible photons.

When a UV photon is converted to a visible photon, what percentage of the UV energy is converted to visible light energy?

A. 20%
B. 28%
B. 36%
D. 55%

The fraction converted is the ratio of the energy of a visible photon to the energy of an ultraviolet photon. Therefore, we have:

$$E_{vis}/E_{UV} = (hc/\lambda_{vis})/(hc/\lambda_{UV}) = \lambda_{UV}/\lambda_{vis} = (200 \text{ nm})/(550 \text{ nm}) = 0.36 = 36\%$$

In a vacuum, light travels at 3×10^8 m/s, but it travels slower when moving through material mediums like air, water, or glass. (In air, the speed of light is only slightly less than it is in a vacuum. This small difference in speed is usually ignored.) When a light wave (or any type of wave) strikes an interface between two different materials where the wave speed *changes*, a portion of the wave will be reflected even if the material is transparent to the wave.

For various materials, we define a quantity called the **index of refraction**. It is the ratio of the speed of light in a vacuum to the speed of light in the material. Since the speed of light is greatest in a vacuum, the index of refraction for every substance is some number *greater* than one. If v is the speed of light in a particular material and c is the speed of light in a vacuum, the index of refraction (n) of the material is defined to be:

$$n = c/v \qquad\qquad \textit{index of refraction}$$

Example 8.2

Light with a wavelength of 450 nm in water enters a piece of glass. If the index of refraction of the glass is 1.5 and the index of refraction of the water is 1.33 find the wavelength of the light in the glass.

The ratio of the index of water to the index of glass is:
$$n_w/n_g = (c/v_w)/(c/v_g) = v_g/v_w$$
The frequency does not change so the ratio of the velocities is:
$$n_w/n_g = v_g/v_w = (\lambda_g f)/(\lambda_w f) = \lambda_g/\lambda_w$$
Solving for the wavelength in glass we have:
$$\lambda_g = \lambda_w(n_w/n_g) = (450 \text{ nm})(1.33/1.5) = 400 \text{ nm}$$

Many of the phenomena that we studied with sound waves also apply to electromagnetic waves. For example, electromagnetic waves will be Doppler shifted if the source and observer are moving relative to each other, and we observe various types of interference with light waves. Light does display some unique qualities, however, do to the fact that is does not need a medium to travel in and its energy is quantized in photons. The Doppler effect formula for light is different than the formula for sound *except* in situations where the relative speeds are *small* compared to the speed of the wave. In this slow speed case, the fractional change in frequency (or wavelength) is proportional to the relative speed of the observer and source.

$$\frac{\Delta f}{f} = \frac{\Delta \lambda}{\lambda} = \frac{v}{c} \qquad\qquad \textit{(Doppler formula for v<<c)}$$

Where Δf is the change in frequency, $\Delta \lambda$ is the change in wavelength, v is the relative velocity of approach or recession of the observer and source, and c is the speed of the wave (either the speed of light or the speed of sound).

Example 8.3

Police radar is based on the Doppler effect. Radar waves with a wavelength of 0.1 m are bounced off a moving vehicle and a frequency shift of 460 Hz is measured. Find the speed of the vehicle. (Remember that there is a double shift since the wave is *reflected* off a moving vehicle.)

Since there is a double shift, $\Delta f = (460 \text{ Hz})/2 = 230 \text{ Hz}$. The frequency and wavelength of the radar wave are related by the expression $f\lambda = c$ (where c is the speed of light). Using the Doppler formula to find the speed of the vehicle we have:

$$v = c(\Delta f/f) = c(\lambda \, \Delta f/c) = \lambda \, \Delta f = (0.1 \text{ m})(230 \text{ Hz}) = 23 \text{ m/s}$$

8.2 Reflection and Mirrors

Light is an electromagnetic wave that travels in straight-line paths, and we often draw straight lines called **rays** to show how light travels from one point to another. When a light ray (or any type of wave) strikes a significantly different surface, some of the light is reflected, some is absorbed by the material, and some of the light may travel through the material if the material is transparent to the wave like glass or water.

The angle the incident ray (incoming ray) makes with the *normal* to the surface is called the **angle of incidence** (θ_1 in Fig. 8.2). The angle the reflected ray makes with the *normal* is called the **angle of reflection** (θ_3), and the angle the refracted ray (the ray that enters the material) makes with the *normal* is called the **angle of refraction** (θ_2).

If a surface is smooth (smooth on a scale compared to the wavelength of the wave), a reflected ray will bounce off the surface so that the angle of incidence and angle of reflection are equal. Both rays also lie in the same plane. This statement is sometimes called the **law of reflection**.

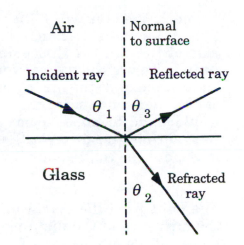

$$\theta_i = \theta_r \qquad \textit{law of reflection}$$

Where θ_i is the angle of incidence and θ_r is the angle of reflection.

When you see light from an object that has been reflected off a plane or flat mirror, the image you see appears to be behind the mirror. The distance from the mirror to the

Figure 8.2 An incident light ray strikes the air-glass interface.

image (**image distance**) is *equal* to the distance from the mirror to the object (**object distance**). Since the light rays do *not* actually pass through points at the image, we say the image is a **virtual image**. A **real image** is formed when light rays actually passing *through* the image. If a piece of paper is placed at a *real* image, the image will be seen projected on the paper. A real image is formed by the projector in a movie theater. If you look into a mirror or through a lens and see an image it *must* be a virtual image. Real images can only be seen if they fall on something like a piece of paper or a screen.

A reflective surface can be curved. For example, a mirror could be made from a *portion* of a reflective sphere. (In most instances, curved reflective surfaces are portions of spherical surfaces since spherical shapes are the easiest to make.) If the spherical mirror curves out, like the side view mirrors on many cars, we say the mirror is **convex**. If the spherical mirror curves in, like a shaving mirror, we say the mirror is **concave** (it is caved in).

If you look into a convex mirror you will see a virtual image that is **upright** (right side up), but is smaller than the object. If you look into a concave mirror that you are very close to, you will see an upright virtual image that is larger than the object.

Generally spherical mirrors are sections of a sphere, and the center *on* the mirror is called the vertex. The symmetrical line drawn from the vertex through the center of the spherical surface is called the **optic axis**. See the figure to the right.

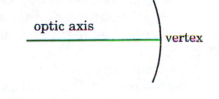

A concave mirror can produce a *real* image. If light from a distant object (like a star) hits a concave mirror, the light will be reflected back to a point and a real image of the star will be formed (Fig. 8.3). If the light from the star comes in parallel to the optic axis of the mirror, this point is called the **focal point**. To see this image you would have to put a piece of paper at the image, or you could put a piece of film there and record the image on the film. (You would have to make a box to hold the film and control the exposure, but the box would *not* need a lens since the concave mirror would be serving as a lens.)

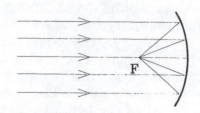

Figure 8.3 Light from a distant point converges on the focal point (F) of the concave mirror.

The light from the star would be parallel as it hit the mirror, and the distance from the mirror to the image is called the focal length of the mirror. The **focal length** of a mirror or lens is the distance from the mirror or lens to the image when the object is an infinite distance away. (An infinite distance is defined to be a very large distance compared to the focal length.) Light parallel to the optic axis is brought to focus at the focal point which is located one focal length from the mirror or lens.

The focal length of a spherical mirror is one half the radius of curvature of the spherical surface. There is a simple relationship between the focal length of the mirror (f), the image distance (d_i), and the object distance (d_o).

$$\frac{1}{d_o} + \frac{1}{d_i} = \frac{1}{f} = \frac{2}{R} \qquad\qquad mirror\ equation$$

Where R is the radius of curvature of the mirror. When making calculations using the mirror equation, the focal length and radius of curvature of a *concave* mirror are considered to be a *positive* quantities, but the focal length and radius of curvature of a *convex* mirror are *negative*.

For single mirrors, the object distance is *positive*. The image distance is positive if the image is on the *same* side of the mirror as the object, but it is negative if the image is on the *opposite* side. A negative image distance means that the image is a *virtual* image (like it is for a plane mirror). The equation can also be used for a plane mirror if we realize that a plane mirror has an *infinite* radius of curvature. The equation tells us that the image distance is the negative of the object distance. The negative sign tells us that the image is behind the mirror and, therefore, must be a virtual image since no light from an object can get *behind* a mirror.

Remember that incident light parallel to the optic axis of a concave mirror will be focused at the focal point. The mirror equation verifies this statement. As the object distance approaches infinity (the incident rays from the object will then be parallel), the image distance approaches the focal length. On the other hand, if an object is placed at the focal point, the light from the object will leave the mirror parallel to the optic axis (the image distance will be infinite).

Spotlights are concave mirrors with a bright flame located at the focal point of the mirror. The light from the flame bounces off the mirror and leaves parallel to the optic axis of the mirror.

The size of the image compared to the size of the object is called the **lateral magnification** of the mirror. If we let the height of the object be h_o and the height of the image be h_i, then the lateral magnification (m) is defined to be:

$$m = \frac{h_i}{h_o} = -\frac{d_i}{d_o}$$ *lateral magnification*

This relationship between the heights and the image and object distances can be verified using similar triangles (Fig 8.4). Most physics texts include the negative sign to indicate the *orientation* of the image. A positive magnification means the image is *upright*, and a negative magnification signifies an *inverted* image. The height of an inverted image is considered to be a *negative* quantity.

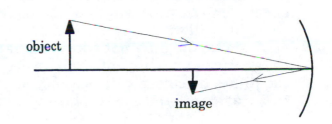

Figure 8.4 Using similar triangles, we see that the ratio of the image height to the object height is equal to the ratio of the image distance to the object distance.

Example 8.4

An object is placed 10 cm from a concave mirror with a radius of curvature of 4 cm. Determine where the image is located and tell if it is real or virtual, inverted or upright.

The focal length is 2 cm (half the radius of curvature). Using the mirror equation $\left(\dfrac{1}{d_o} + \dfrac{1}{d_i} = \dfrac{1}{f} \right)$ we have:

$1/10 + 1/d_i = 1/2$
$1/d_i = 1/2 - 1/10 = 4/10$
$d_i = 2.5$ cm

The image is real since the answer is positive. The image is also inverted since the magnification is negative. $m = -(d_i)/(d_o) = -(2.5)/10 = -0.25$

8.3 Refraction and Total Internal Reflection

If a light wave strikes the interface between two different materials at an angle, the change in the wave's speed at the interface will cause the wave entering the second material to change its direction of travel. This effect is called **refraction**. Snell's law gives a relationship between the angle of refraction, the angle of incidence, and the indices of refraction of the two materials (Fig. 8.2).

$$n_1 \sin\theta_1 = n_2 \sin\theta_2 \qquad\qquad Snell's\ law$$

The material with the larger index of refraction is said to be more *optically* dense than the other material. If a light ray enters a medium that is *more* optically dense, the ray will bend *toward* the normal to the surface.

The index of refraction generally depends slightly on the wavelength of the light. If white light (light of all colors) enters glass at an angle, the different wavelengths will be refracted at slightly different angles and the different colors will be separated as they travel into the glass. This separation of light into its different colors is called **dispersion**. *Shorter* wavelengths are refracted *more* than longer wavelengths. Light passing through a prism will be

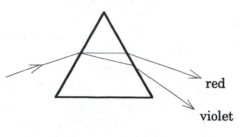
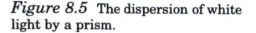

Figure 8.5 The dispersion of white light by a prism.

dispersed as it enters and *again* as it leaves the prism, producing a spectrum. (Fig. 8.5)

If a ray is incident on a *less* optically dense medium, the ray will bend *away* from the normal. (An example would be a ray of light traveling upwards through the water in a lake and then hitting the surface where the water meets the air.) At some critical incident angle, the ray that is refracted into the air will have an angle of refraction of 90°. For incident angles *greater* than this critical angle, no light is refracted and *all* the incident light is reflected off the interface. This effect is called **total internal reflection**. This 100% reflection is called *internal* reflection because it can only occur when the incident light ray is *inside* a medium that is *more* optically dense than the medium beyond the boundary.

Total internal reflection is utilized in many optical instruments. Prisms are used instead of mirrors in binoculars because a prism utilizing total internal reflection can reflect 100% of the light, but even the best mirrors can only reflect about 95% of the incident light. In typical binoculars, the light is reflected four times and if 5% was lost at each reflection, only about 81% of the entering light would get to your eye. Total internal reflection is also utilized in fiber optics. If a light ray enters the end of a solid optical fiber, the entire ray will be reflected off the sides as it travels along the fiber. No light is lost out the sides of the fiber. Endoscopes are instruments used to view internal portions of the body. They use fiber optics to transmit the light needed to illuminate the object being viewed, and fiber optics to transmit images to the outside.

Example 8.5

Light enters a prism and is reflected as shown. What
is the minimum index of refraction of the glass if 100%
of the light is to be reflected?
A. 1.0
B. 1.2
C. 1.4
D. 1.5

Since the angle of incidence is 45° at the *two* places where the wave
reflects off walls of the prism, and the angle of refraction must be 90° for
total internal reflection, Snell's law ($n_1 \sin\theta_1 = n_2 \sin\theta_2$) gives:

$n \sin(45°) = 1 \sin(90°)$ Where n is the index of refraction of the glass.
Solving for *n* we obtain:

$n = 1/(\sin 45°) = 1/(1/\sqrt{2}) = \sqrt{2} = 1.4$ The answer is C.

8.4 Lenses

Most lenses are made of transparent material like plastic or glass. The two
surfaces of a lens are generally ground to a spherical contour since spherical
surfaces are the easiest to grind. A **converging lens** has a thick center and
thinner edges. (Although technically incorrect, some people call these lenses
convex lenses.) The symmetrical axis through the center of the lens is called
the optic axis. Converging lenses are so named because incident light that is
parallel to the optic axis of the lens will converge at a focal point on the
opposite side of the lens. The focal length of the lens is the distance from the
lens to the focal point. If the parallel light came from a distant star, a *real*
image of the star would be formed at the focal point of the lens.

A magnifying glass is a converging lens that is used to form a magnified *virtual*
image as you look at an object through the lens. To form this virtual image,
the object must be located *closer* to the lens than the focal length of the lens.
Most of us have used a converging lens like a magnifying glass to form a *real*
image of the Sun. The image is so bright that a piece of paper can be burned
by the Sun's image.

A camera lens is a converging lens that forms a real image on the film when
the camera's shutter is opened. A movie projector lens is also a converging
lens that forms a real image on the movie screen. The movie film is the object.
For a converging lens to form a real image, the object must be located *further*
from the lens than its focal length.

The *strength* of eyeglasses depends on the focal length, but opticians specify
the strength as the *reciprocal* of the focal length measured in *meters*. The unit

of lens power is the **diopter**. One diopter (abbreviated D) is equal to one *inverse* meter. The lens power is given by the following expression:

$$P = \frac{1}{f}$$
                                                            lens power

Where P is the lens power and f is the focal length of the lens in meters.

A **diverging lens** has a thin center and thicker edges. (Although technically incorrect, some people call diverging lenses concave lenses.) Incident light parallel to the optic axis of a diverging lens will spread out or diverge after it passes through the lens and will *appear* to be coming from a focal point on the *same* side of the lens as the incident light. Diverging lenses *cannot* form a real image of a real object. If you try to use a diverging lens the way you would use a magnifying glass, you will see a virtual image that is smaller than the object.

A nearsighted person needs glasses with diverging lenses to correct their problem. The eyes of a nearsighted person focus the light *in front of* the retina, but the diverging lenses spread the light out slightly so it focuses on the retina.

A far sighted person needs converging lenses to correct their vision. Their eyes focus the light *behind* the retina and the converging lenses converge the light enough to focus it on the retina.

The relationship between the object distance (d_o), the image distance (d_i), and the focal length (f) of a lens is identical to the equation for mirrors:

$$\frac{1}{d_o} + \frac{1}{d_i} = \frac{1}{f}$$
                                                            thin lens equation

This equation is called the **thin lens equation** because it only applies to relatively thin lenses. To solve any problem using this equation, a positive or negative sign must be included with each quantity. Focal lengths of converging lenses are taken as *positive*, but focal lengths of diverging lenses are *negative*. (Since the focal length of a diverging lens is negative, the *power* of a diverging lens is also negative.) Object distances are positive if a single lens is involved. Image distances are positive if the image is on the *opposite* side of the lens from the object and negative if they are on the *same* side. Negative image distances imply that the image is *virtual*.

The *lateral* magnification of a lens is given by the same equation we used for mirrors. When a converging lens is used as a magnifier, the **angular magnification** or **magnifying power** of the lens is used to describe the magnification of an object viewed through the lens. The angular magnification is a measure of how much the *angular size* of the object has been increased by the magnifying glass. The angle subtended by an object being viewed by a 10x magnifier would be increased by a factor of 10.

8.5 Diffraction

When a water wave hits a wall with a small opening, the wave that passes through the opening spreads out in all directions. This spreading effect is known as **diffraction**. Diffraction of water waves can be easily understood if we imagine a water wave hitting a wall that contains a small opening. As a crest hits the wall the water in the opening rises, and as a trough hits the wall the water in the opening lowers. The water in the opening rises and lowers periodically as the waves strike the wall.

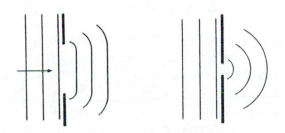

Figure 8.6 Diffraction through a large opening (left) and through a small opening.

the wall. The water in the opening behaves as if someone were wiggling the water with a stick, and circular waves travel out away from the opening into the water beyond the wall. We say the wave is diffracted by the opening.

In a similar fashion light waves are diffracted as they pass through small openings. The smaller the opening, the more diffraction. If the opening is small compared to the wavelength of the wave, the wave will spread out in all directions, but if the opening is much larger than the wavelength of the wave, only the wave near the edges of the opening will be diffracted (Fig. 8.6).

Diffraction of waves imposes limitations on the ability of optical instruments to distinguish detail. Waves entering the objective of a microscope or the front end of a telescope will be diffracted, and the image formed will be deformed because of the diffraction caused by the opening of the optical instrument. The *larger* the opening, the *less* the distortion will be.

A measure of the clarity of an image formed by a telescope (or any instrument that uses waves to create an image) is the **resolution** of the instrument. The smallest detail the instrument is able to distinguish is a measure of the resolution of the instrument. Resolution is usually expressed as the smallest *angular* detail that can be distinguished. For example, the resolution of the human eye is about one minute of arc. (The resolution of the human eye is limited by how closely the rods and cones are packed, but the size of the pupil limits the theoretical resolving power of the eye.) Jupiter is a round ball, but when we look at it with our naked eyes, all we see is a point of light. We see a dot because the *angular size* of Jupiter is less than one minute of arc (less than the resolution of the human eye). In order to see a round ball, a telescope must increase the angular size of Jupiter to a value larger than one minute of arc.

The resolution of an instrument using waves to form an image can be *increased* (it will be able to detect smaller details) if *shorter* wavelength waves are used. For example, if a certain wavelength sound wave is used in an ultrasound exam, the detail seen could be increased by using a sound wave with a *shorter* wavelength (higher frequency). If an image is formed from the waves reflected off an object, the size of the smallest detail that can be resolved is limited to the wavelength of the wave used.

Resolution is quite different from magnification. A picture can be magnified by any amount, but beyond a certain point no more *detail* will be distinguished. If a picture lacks detail, no amount of magnification will help. Resolution is a measure of the *clarity* of the picture. A high resolution photograph will show more detail than a low resolution photograph even if both are magnified by the same amount. A telescope can be made with any magnification you desire, but you will still not be able to resolve details on the surface of a nearby star because the telescope does not have the necessary resolution. Useful magnification is limited by the resolution of the optical instrument.

Moving electrons behave like waves. They have a wavelength that is *inversely* proportional to the momentum (or speed) of the electrons. Fast moving electrons have a *shorter* wavelength than slower moving electrons. Moving electrons are the "waves" used in an electron microscope. The wavelength of an electron that has been accelerated through 1000 V is about 0.039 nm, much shorter than visible light which has wavelengths between 400 nm and 750 nm. The useful magnification of a visible light microscope is less than 2000X whereas the useful magnification of an electron microscope is over 100,000X.

Example 8.6

Ultrasound waves traveling at a speed of 1500 m/s in the human body will reflect off various objects such as babies. The ultrasound reflections are electronically constructed and displayed on a monitor. If the smallest detail that must be resolved is 10^{-3} m, what is the lowest wave frequency that can be used?

The size of the smallest object that can be resolved is approximately equal to the wavelength of the waves used, therefore, the wavelength must be shorter than 10^{-3} m. The frequency times the wavelength gives the wave speed, so the lowest frequency that can be used is:

$f = v/\lambda = (1500 \text{ m/s})/(10^{-3} \text{ m}) = 1.5 \times 10^6$ Hz

8.6 Aberrations

In optical instruments, lenses and mirrors are used to form images. However, it is impossible to design a lens or mirror that forms a perfect image. All lenses are shaped somewhat like prisms so they disperse the light slightly. Therefore, the colors of light are not focused at exactly the same point. This problem is known as **chromatic aberration**, and in extreme cases it results in rainbows surrounding all the images. A spherical concave mirror will not even focus all parallel light rays at the same focal point, a problem called **spherical aberration**. Other aberrations include astigmatism, coma, pin cushion distortion, and barrel distortion. These aberrations are the imperfections found in most optical systems. Even if we ignore diffraction effects, it is impossible to design an optical system that will form a perfect image by exactly focusing all the rays that pass through the system.

8.7 Polarized light

Light is an electromagnetic wave and its electric and magnetic fields oscillate in a direction that is *perpendicular* to the direction of travel of the wave. Therefore, electromagnetic waves are *transverse* waves. If a light wave is coming toward you, the electric field may be wiggling vertically (up and down) or it could be wiggling horizontally (side to side). Since the electric field is a vector, an electric field of any orientation can be resolved into a vertical and horizontal component so we will just discuss these two orientations. The light waves coming from an ordinary light bulb have waves with all possible orientations. That is, vertical and horizontal orientations are equally likely. It is possible to produce light where all the electric field vectors are oriented in the same direction. Such light is called **polarized light**. Certain substances (called polarizers) have the ability to remove all the waves with one orientation, so the light emerging from the material is polarized.

Light *reflected* off smooth surfaces (like water) tends to be highly polarized since the horizontally oriented waves are reflected more easily than the vertically oriented waves. Also, light scattered by air molecules is highly polarized.

Questions and Problems

1. Which of the following best explains why ultraviolet light does not pass through Earth's atmosphere.
   A. UV light has less energy than visible light and cannot penetrate Earth's atmosphere.
   B. UV light has such a low frequency that it is easily absorbed by the air.
   C. UV light has a shorter wavelength than visible light and is easily absorbed by the air.
   D. UV is reflected by Earth's upper atmosphere.

2. What information is needed to determine the speed of light in water?
   A. The wavelength of the light in water and the wavelength in air.
   B. The speed of light in a vacuum and the index of refraction of water.
   C. The speed of light in a vacuum and the wavelength in water.
   D. The speed of light in a vacuum and the density of water.

3. Ultrasound reflects off different objects within the human body because:
   A. The waves are reflected at boundaries where its frequency changes.
   B. The waves are reflected because their speed changes at the boundary.
   C. The waves reflect off objects with different shapes.
   D. The waves loose energy as they pass through the body.

4. A concave mirror is made from a section of a reflecting sphere with a radius of 50 cm. If an object is placed 50 cm from the mirror, how far from the lens will the image be formed?

5. The concave objective mirror in the Lick Observatory telescope has a diameter of 3 meters and a focal length of 52.5 meters. The angular magnification of a telescope is equal to the focal length of the objective lens or mirror divided by the focal length of the eyepiece. If an eyepiece with a focal length of 25 mm is used on the Lick telescope, what is the angular magnification?

6. A large concave mirror is the primary light collector of an astronomical telescope. The image of the Moon formed by such a mirror will be:
   A. Real and inverted
   B. Virtual and inverted
   C. Real and upright
   D. Virtual and upright

7. You look into an optical device and see a virtual image. How can you tell it is a virtual image?
   A. Real images are sharper than virtual images.
   B. Real images are inverted.
   C. Real images are brighter than virtual images.
   D. A virtual image will move as you move your eye around.

120

8. A laser beam hits a plane mirror and the reflected beam travels back along the same path as the incident beam. If the plane mirror is rotated through an angle of 20°, the angle between the reflected and the incident ray will be:
   A. 10°
   B. 20°
   C. 30°
   D. 40°

9. A converging lens has a focal length of f. If an object is placed a distance of $3f$ from the lens, an image will be formed a distance of $1.5f$ from the lens. The ratio of the height of the image to the height of the object is:
   A. 2
   B. 1/2
   C. 1.5
   D. 3

10. Microwaves with a frequency of 4×10^{10} Hz reflect off a car and are measured to be 2×10^3 Hz lower than the transmitted signal. What could explain this change in frequency?
    A. The car is moving toward the microwave transmitter.
    B. The car is moving away from the microwave transmitter.
    C. The car's engine is running and is interfering with the microwaves.
    D. The distance to the car is large and the frequency decreases with time.

11. While lying on the bottom of a lake, a diver looks straight up at the surface of the calm lake and sees the blue sky, but a little further off to the side he sees the bottom of the lake. What causes this strange effect?
    A. This could not happen.
    B. Light from the sky will be reflected off the surface of the lake.
    C. This effect is due to the wide angle vision of the face mask.
    D. If the angle of incidence is large enough, light from the bottom will be totally internally reflected at the surface.

12. The particle nature of light is best demonstrated by which of the following?
    A. Light can be polarized.
    B. Light exhibits diffraction.
    C. The wavelength of light determines the color of the light.
    D. The energy of light is quantized.

13. If an X-ray photon has an energy of 6.0×10^{-14} J, what is its wavelength? Look up any constants you need, but do not use a calculator.

14. Sound waves cannot be polarized because:
    A. Sound waves cannot be diffracted.
    B. Sound waves do not display interference.
    C. Sound waves are longitudinal waves.
    D. Sound waves are transverse waves.

Answers to Questions and Problems

1. A. Ultraviolet light (a photon) has *more* energy than visible light.
    B. Ultraviolet light has a *high* frequency compared to visible light.
    C. This is the correct answer.
    D. Ultraviolet light is *not* reflected by air.

2. A. This information will only allow us to calculate the *ratio* of the speed of light in air to the speed of light in a vacuum.
    B. The index of refraction is the ratio of the speed of light in a vacuum to the speed of light in air. This answer is the correct one.
    C. The frequency must also be known to get the speed in water.
    D. The density of a substance is not related to the speed of light in that material.

3. A. The frequency does not change at a boundary. The number of waves leaving one region per second is the same as the number of waves entering the next region per second.
    B. Reflection always occurs at a boundary where the speed changes. This is the correct answer.
    C. The shape has nothing to do with reflection although if the speed changes, reflection will occur.
    D. Waves may loose energy when passing through material, but reflection does not necessarily accompany this loss of energy.

4. The focal length is 25 cm (one half the radius of curvature). The image distance can be obtained from the mirror equation: $\dfrac{1}{d_o} + \dfrac{1}{d_i} = \dfrac{1}{f}$

    $$\frac{1}{d_o} + \frac{1}{50 \text{ cm}} = \frac{1}{25 \text{ cm}}$$
    $$\frac{1}{d_o} = \frac{1}{50 \text{ cm}}$$
    $$d_o = 50 \text{ cm}$$

5. The angular magnification is given by: $(52.5 \text{ m})/(25 \times 10^{-3} \text{ m}) = 2.1 \times 10^3$

6. Concave mirrors produce real inverted images if the object is further away than the focal length of the mirror. The answer is A.

7. Real images are not necessarily sharper, inverted, or brighter. The answer is D. Look at a virtual image in a mirror and move around. The virtual image will move around as you move. As you move around in a movie theater, however, the real image on the screen does not move.

8. When the plane mirror is rotated 20° the normal line is also rotated 20°, so the angle of incidence is 20°. The angle of reflection must also be 20°, so the reflected ray makes a 40° angle with the incident ray. The answer is D.

9. The ratio of the height of the image to the height of the object is equal to the ratio of the image distance to the object distance. This ratio is 1/2. The answer is B.

10. The Doppler effect causes a shift to lower frequency when the source and/or observer are moving apart. The answer is B.

11. The answer is D. Answer B is correct, but does not explain the effect described.

12. A. True, but demonstrates that light is a *transverse* wave.
    B. This statement is true, but demonstrates that light is a *wave*.
    C. This statement is true, but does not support the particle nature of light.
    D. The fact that photons carry a definite quantity of energy demonstrates the particle nature of light. The answer is D.

13. The energy and wavelength are related by: $E = hc/\lambda$. Therefore, the wavelength is:
    $$\lambda = hc/E = (6.6 \times 10^{-34} \text{ J·s})(3 \times 10^8 \text{ m/s})/(6.0 \times 10^{-14} \text{ J}) = 3.3 \times 10^{-10} \text{ m}$$

14. Only transverse waves can be polarized and sound waves are longitudinal waves. The answer is C. All waves can be diffracted and all display interference.

9 Atomic and Nuclear Physics

9.1 Photoelectric Effect

When light hits a metal surface, electrons may be emitted from the surface. This phenomenon is known as the **photoelectric effect** since photons initiate the release of electricity in the form of electrons. Whether electrons are emitted depends *not* on how bright or intense the light is, but on the frequency of the light. The energy of a photon is proportional to its frequency, and if the energy of each photon is not large enough, no electrons will be emitted.

The electrons in the metal are bound by electrical forces, and it takes a certain minimum energy to liberate electrons from the metal. This *minimum* energy is called the **work function** of the metal. The work function depends on the composition of the metal. When a photon is absorbed by an electron in the metal, some of the photon's energy is used by the electron to escape from the metal and the remainder goes into kinetic energy of the free electron. This relationship can be written as:

$$hf = \text{KE}_{max} + W_o$$

Where hf is the energy of the photon, W_o is the work function of the metal, and KE_{max} is the *maximum* kinetic energy of the emitted electrons. Some electrons are more tightly bound by the metal and require more than the minimum energy (W_o) to escape the metal. These electrons are emitted with *less* kinetic energy than the maximum. The photoelectric effect can only be understood if we realize that light energy is quantized. In the photoelectric effect, the light energy is absorbed as individual quanta or photons.

Example 9.1

A certain metal has a work function of 2.9 eV. When monochromatic (all of one wavelength) ultraviolet light shines on the metal, electrons are released. The most energetic electrons can be stopped by an electric potential difference of 1.2 V. The energy of an ultraviolet photon is:
A. 1.2 eV
B. 1.7 eV
C. 2.9 eV
D. 4.1 eV

The most energetic electron can be stopped by an electric potential difference of 1.2 V, therefore, it must have had a kinetic energy of 1.2 eV. The energy of an UV photon must be used to free that electron from the metal and supply it with this kinetic energy. Therefore, the energy of the photon must be: $E = 2.9 \text{ eV} + 1.2 \text{ eV} = 4.1 \text{ eV}$. The answer is D.

9.2 Blackbody Radiation

Solids and liquids emit electromagnetic radiation with a very distinctive spectrum known as a blackbody spectrum. A spectrum is a plot of the intensity of the radiation as a function of the wavelength (or frequency). A **blackbody spectrum** is the spectrum produced by an ideal radiator, but most solids and liquids are approximately ideal radiators. An ideal radiator is also an ideal absorber since an object in equilibrium (one whose temperature is constant) will emit energy at the same rate at which it absorbs energy. Near room temperature, ideal absorbers are black, which is where the name blackbody comes from.

As the temperature of a blackbody increases, the *total* energy it emits over all wavelengths per second increases as the fourth power of the absolute temperature (Fig. 9.1). This relationship is known as **Stefan's law**. When the temperature of an object increases beyond about 1000 K (about 700°C) it begins to glow red since a significant fraction of the radiation emitted is in the visible portion of the spectrum. For objects hot enough to glow, Stefan's law says that as an object gets hotter it gets *brighter*.

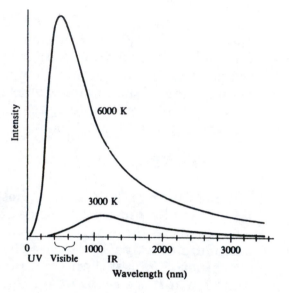

Figure 9.1 Blackbody spectrum at two different temperatures. The total energy emitted (area under the curve) increases as the fourth power of the absolute temperature.

Another important property of blackbody radiation is that the *wavelength* at which most of the energy is emitted (peak of the curve in Fig. 9.1) *decreases* as the temperature *increases*. This relationship is known as :

$$\lambda_p \, T = 2.90 \times 10^{-3} \text{ m·K} \qquad\qquad \textit{Wien's law}$$

Where T is the absolute temperature and λ_p is the wavelength at which most of the energy is emitted. Wien's law explains why a blackbody that is hot enough to glow changes *color* as their temperature changes. A hot superdense gas like we find inside stars also produces a blackbody spectrum, and the color of the star is related to the temperature by Wien's law. The coolest stars are red. Orange stars are hotter than red stars, and yellow stars are even hotter. A blackbody whose energy peaks in the yellow-green region of the spectrum looks *white* to our eyes, and the hottest stars look bluish.

126

Example 9.2

The Sun is a 1.4×10^9 m diameter yellow star whose peak energy is at about 550 nm. What is the temperature of the Sun's surface?

The temperature is obtained from Wien's law (the diameter is not needed):

$$T = (2.9 \times 10^{-3} \text{ m·K})/\lambda_p = (2.9 \times 10^{-3} \text{ m·K})/(550 \times 10^{-9} \text{ m}) = 5300 \text{ K}$$

9.3 Energy levels of an Atom

Solids, liquids, and very dense gasses produce blackbody type spectra. They emit photons of all wavelengths (colors), and we call this type of spectrum a **continuous spectrum**. The light given off by these substances is actually given off by the electrons in the material. The electrons oscillate and emit electromagnetic waves (photons). The atoms in these materials are very close together and the electrons are influenced by the neighboring atoms and electrons. The electrons are free to oscillate at *any* frequency and, therefore, photons of *all* wavelengths are emitted.

The atoms in a gas, however, are separated by great distances and except for an occasional collision, the electrons are not greatly influenced by their neighboring atoms. For isolated atoms, the electrons can only orbit their respective nuclei on well defined paths that we call orbits. Since the electrons are restricted to specific orbits, they have a specific amount of energy.

An electron can jump to an orbit with a different energy, but in order to make the transition it must acquire or lose an amount of energy equal to the *difference* in energy of the two orbits. For example, if an electron is in an orbit with an energy of -3 eV (energies of bound electrons are negative) and it jumps to an orbit with an energy of -5 eV, it must give up 2 eV of energy. It will generally emit this 2 eV of energy as a photon.

If an electron is in the -5 eV energy orbit, it could absorb a photon with 2 eV of energy (if such a photon happens to hit it) and jump to the -3 eV orbit. Another way the electron could get the 2 eV of energy is in a *collision* with another fast moving atom. In hot gasses, the atoms are moving very fast and when two atoms collide, an electron can use some of the kinetic energy in the collision to jump up to a higher energy orbit or **energy level**. If an atom has an electron in a higher energy level we say the atom is **excited**. An atom can be excited either by absorbing a photon with just the right energy, or it can be collisionally excited when it collides with another atom.

Atoms do not stay in an excited state very long. They will generally emit photons as the electron jumps to lower energy levels. When the electron reaches its lowest energy level, we say it is in the **ground state**. Hot gasses like those in a fluorescent light give off light because the atoms are being collisionally excited (by collisions with the high energy electrons that are

127

running through the tube), and the excited atoms then emit photons as their electrons jump down to lower energy levels.

Hot gasses do not emit all wavelengths like the atoms in a solid or liquid. They emit specific wavelengths corresponding to the *differences* in the energy levels of the atoms in the gas. The type of spectrum produced by a gas is called an **emission** or **bright-line spectrum** since only certain wavelengths (called emission lines) are present. Every type of gas produces its own unique emission spectrum, and the gas can be identified from the emission lines present in its spectrum. The intensity of an emission line depends on the *probability* of the transition that produces the line. The more probable the transition is, the more intense the line will be.

Example 9.3

A certain gas atom has the following energy levels:
Level 4 -0.3 eV
Level 3 -0.8 eV
Level 2 -1.5 eV
Level 1 -4.2 eV

1. What is the maximum number of emission lines in the spectrum?
2. If visible light photons have energies between 1.8 eV and 3.0 eV, how many of the emission lines are in the visible region of the spectrum?

1. There are six possible transitions: $4\rightarrow1$, $3\rightarrow1$, $2\rightarrow1$, $4\rightarrow2$, $3\rightarrow2$, and $4\rightarrow3$. Therefore, there are six emission lines in the spectrum.
2. The energies of the six transitions in question one are: 3.9 eV, 3.4 eV, 2.7 eV, 1.2 eV, 0.7 eV, and 0.5 eV.
   The 2.7 eV photon is the only one in the visible region of the spectrum. The 0.5 eV, 0.7 eV, and 1.2 eV photons are in the infrared region, and the 3.9 eV and 3.4 eV photons are ultraviolet photons.

9.4 Nuclear Reactions and Radioactivity

The nuclei of all atoms except the common form of hydrogen are made of two types of particles, positively charged **protons** and electrically neutral **neutrons**. Protons and neutrons are collectively referred to as **nucleons**. Atoms with the same number of protons and, therefore, the same *chemical* properties, are given the same **element** name. For example, any atom with two protons in its nucleus has the chemical properties we associate with the element helium. The most common form of helium also has two neutrons, but a nucleus with two protons and one neutron is still helium, although it is a different form that we call helium-3, which is also written $^3_2$He. The upper number is the atomic **mass number** which is the number of nucleons (protons plus neutrons) and the lower number is the **atomic number** which is the number of protons. Different forms of the same element are called **isotopes** of that element. There can be many isotopes of a given element, differing only by the number of neutrons they possess.

The *actual* mass of a nucleus is measured in units called **unified atomic mass units** (abbreviated u). One atomic mass unit is equal to 1.66×10^{-27} kg. The mass of a nucleus measured in atomic mass units is *approximately* equal to the atomic mass number. The two quantities are not exactly equal because the mass changes slightly with the binding energy of the nucleus.

The nucleons in a nucleus attract each other because of a very *short-range* force called the nuclear force or the strong interaction. Only nucleons feel the strong interaction; electrons do not. To feel this strong attractive force, the nucleons have to be very close to each other. The positively charged protons also repel each other because of the longer range electromagnetic force, but since the nuclear force is stronger at short distances, most nuclei are held together very tightly.

However, if a nucleus is too large, it may not be completely stable. Because of the short-range nature of the nuclear force, nucleons on opposite sides of a large nucleus may be so far apart that their nuclear attraction is greatly reduced. Since some of these nucleons are protons, their electrical repulsion may be almost as strong as their nuclear attraction. Under these conditions the nucleus can be unstable, and if it is it will eventually change into a different nucleus. This size constraint explains why we find no stable nuclei larger than $^{209}_{83}\text{Bi}$ (bismuth), a nucleus with 209 nucleons (83 protons and 126 neutrons).

Unstable nuclei will emit various kinds of particles until they become stable. A radioactive nucleus is generally referred to as a **parent** nucleus, and the nucleus that is left after the radioactive decay is called the **daughter** nucleus. Unstable nuclei are said to be **radioactive**, and the particles they emit are called radiation. The general use of the word radiation is somewhat misleading because this kind of radiation is composed of small bits of matter and is quite different from the electromagnetic radiation we discussed earlier. Electromagnetic radiation is not composed of material particles, but consists of bundles of *pure* energy in the form of waves.

Some radioactive nuclei, especially very large nuclei like uranium and plutonium, emit helium nuclei (2 protons and 2 neutrons) as a way of becoming smaller. A helium nucleus is emitted instead of a single proton or neutron because the helium nucleus is a very tightly bound object. A fast moving helium nucleus is called an **alpha particle**, and nuclei that emit alpha particles are said to be emitting alpha radiation. Although they do not penetrate very deeply into human tissue, alpha particles are harmful to biological tissues because of the high speed with which they are emitted. (They have a lot of kinetic energy.)

In smaller nuclei, the number of neutrons is approximately equal to the number of protons, and in larger nuclei the percentage of neutrons is greater. Nuclei that are unstable because they do not have the right *ratio* of protons to neutrons can convert a proton into a neutron or vice versa depending on which is needed.

Electric charge is one of those quantities in physics that is conserved. **Conservation of electric charge** means that in any process, no *net* electric charge can be produced or destroyed. When a neutron is converted into a proton, an electron is also created to conserve electrical charge. The reaction is:

$$n \rightarrow p + e^- + \bar{v}$$
neutron decay

The third particle created in the reaction is called a neutrino (actually an antineutrino as represented by the *bar* over the symbol), but it can be ignored for our purposes. The fast moving electron emitted from the nucleus in this reaction is called a **beta particle**, a beta ray, or beta radiation. *Inside* certain radioactive nuclei it is also possible for a proton to change into a neutron. The process that occurs *inside* the nucleus is:

$$p \rightarrow n + e^+ + v$$

The e^+ is a positive particle that is produced to conserve electric charge. It is called a positron, and it is the antiparticle of the electron. When it combines with an electron the two will annihilate each other. A neutrino (not an antineutrino) is also emitted in this reaction.

Emitted electrons are commonly called beta particles, but positrons are also beta particles. We distinguish between the two types of beta decay by referring to β^- decay (electrons) and β^+ decay (positrons). Unfortunately, often people refer simply to beta decay when they really mean β^- (electron) decay.

Because of the speed with which the particles are emitted, beta radiation can be quite harmful to the human body and other biological tissue. They are much more penetrating than alpha particles and can cause cancer in interior organs. Beta emitters in the waste products from atomic power plants and in the debris from nuclear explosions account for much of the danger associated with these substances.

Example 9.4

1. A $^{14}_{6}C$ nucleus decays into a $^{14}_{7}N$ nucleus. What particle (or particles) could be emitted?

Since a neutron has changed into a proton, one beta particle must have been emitted.

2. A $^{238}_{92}$U nucleus decays into a $^{234}_{92}$U nucleus. What particle (or particles) could be emitted?

Four nucleons have been lost in the decay, but they are all neutrons. The four neutrons can be made into an alpha particle by first changing two of them into protons. This transformation would require the emission of two electrons (beta particles). Therefore, two beta particles and one alpha particle must be emitted to complete the decay.

If a nucleus has too many protons, a proton may capture one of the electrons orbiting the nucleus and combine with it to become a neutron. This process is called **electron capture**. The electron is generally one from the innermost shell of the atom (called the K shell) and the process is called K-capture. An electron from a higher orbit jumps down to fill the vacancy and because the energy difference is so great, an X-ray is emitted.

After alpha decay, beta decay, or electron capture the nucleus may be left with an excess of energy. This energy is generally so great that it is emitted as a high energy electromagnetic wave called a **gamma ray**. Gamma rays are very penetrating, and they represent a very dangerous form of radiation that is prevalent in nuclear waste and other radioactive substances.

The number of nuclei that decay in a given period of time is proportional to the time interval and the *number* of parent nuclei present. Mathematically this means that the number of nuclei which have not decayed will decrease in time *exponentially*. The time period during which one half of the nuclei are likely to decay is called the **half-life** of that particular nucleus. An equivalent definition of half-life is the time interval in which a nucleus has a 50/50 chance of decaying.

Example 9.5

A radioactive isotope has a half-life of 9 months. After 3 years, what fraction of a sample of this isotope will be left?
A. 1/4
B. 1/8
C. 1/16
D. 1/32

Three years (36 months) is 4 half-lives and one half the sample decays *every* half-life. Therefore, after 4 half-lives the amount left will be:

$(1/2)(1/2)(1/2)(1/2) = (1/2)^4 = 1/16$ The answer is C.

Since the nuclear force is the strongest force in nature, the energy holding a nucleus together may be so large that the mass of a nucleus is significantly different than the sum of its parts. The change in energy is given by the famous equation of Einstein:

$$E = mc^2$$

Where m is the *change* in mass, E is the *change* in energy, and c is the speed of light. Since this equation is most useful in finding the energy given off in nuclear reactions, nuclear mass are often expressed in units of MeV/c^2. The quantities are related as follows:

$$1 \text{ u} = 1.66 \times 10^{-27} \text{ kg} = 931.5 \text{ MeV}/c^2 \qquad \textit{atomic mass unit}$$

Example 9.6

A 3.0 MeV (million electron volt) gamma ray is emitted by a lead nucleus. By how much is the mass of the nucleus reduced. ($1 \text{ eV} = 1.6 \times 10^{-19}$ J)

The energy (in joules) of the gamma ray emitted is:
$E = (3.0 \times 10^6 \text{ eV})(1.6 \times 10^{-19} \text{ J/eV}) = 4.8 \times 10^{-13}$ J
We use $E = mc^2$ to find the change in the mass.
$m = E/c^2 = (4.8 \times 10^{-13} \text{ J})/(3 \times 10^8 \text{ m/s})^2 = 5.3 \times 10^{-30}$ kg

It is possible to change one nucleus into another by adding of subtracting nucleons. Nuclear **fusion** is the process of combining (fusing together) two smaller nuclei to create one larger nucleus. The following nuclear process is an example of nuclear fusion:

$$^2_1H + {}^3_1H \rightarrow {}^4_2He + n \qquad\qquad \textit{D-T reaction}$$

In the above fusion reaction, two isotopes of hydrogen fuse to form helium. (A neutron is given off in the reaction.) The hydrogen with two nucleons is called deuterium and the hydrogen with three nucleons is called tritium. This reaction, known as the D-T reaction, is the basic reaction in *hydrogen* bombs. Various fusion reactions that convert hydrogen into helium are the source of the Sun's energy. Fusion reactions involving smaller nuclei generally produce a great deal of energy. The energy release results in a *loss* of mass. The final products are *less* massive than the original components.

A few very large unstable nuclei will split into two smaller nuclei, a process called nuclear **fission**. Nuclear fission does not occur widely in nature, but it can be initiated in some large nuclei by hitting the nucleus with a neutron. Chain reactions in *atom* bombs are created by initiating fission reactions in a mass of uranium or plutonium. The fission of one nucleus creates two smaller nuclei and two or three free neutrons that can initiate other fission reactions. Once the process begins, the number of reactions increases rapidly leading to a **chain reaction** and the release of huge amounts of energy in a very short period of time. In nuclear reactors, *neutron* absorbing rods are inserted in the fuel to control the rate at which the reactions occur, and keep the fissionable fuel from exploding.

When energy is released in a reaction, the mass of the final products is less than the mass of the initial products. In chemical reactions, the change in mass is generally too small to worry about, but in nuclear reactions the change in mass can be considerable. When two smaller nuclei fuse to form a larger nucleus, the binding energy of the larger nucleus is released in the reaction. This loss of energy shows up as a loss of mass in the reaction. The most tightly bound nuclei occur near iron on the periodic table. The tight binding of elements near the middle of the periodic table explains why energy can be released as very massive nuclei undergo nuclear fission, but energy can also be released as very small nuclei undergo nuclear fusion.

9.5 X-Rays

As we saw with electron capture, if an electron is removed from the K shell of a reasonably large atom, an electron from a higher orbit will jump down to fill the vacancy. The energy difference between the two states is generally so large, that the photon emitted with the transition is an **X-ray** photon. X-rays in X-ray machines are generally produced by knocking electrons out of the K shell of atoms, especially metal atoms like copper or iron. The metal sample is bombarded with high energy electrons that knock out K shell electrons. The metal emits X-rays as the vacant K shells are filled by electrons from higher orbits.

X-rays enable us to seeing internal structures in the human body because they travel relatively easily through animal tissue. They are absorbed by tissue, but various tissues absorb differently. For example, bones absorb X-rays more easily than muscle tissue so the bones can be easily seen as *shadows* on X-ray photographs. Unfortunately, the X-ray photons that are absorbed by the body tissue deposit a lot of energy at the point where they are absorbed. This energy can cause unwanted chemical changes in the DNA that can create a cancer cell.

Questions and Problems

1. An object that radiates like a blackbody appears white when its temperature is about 6000 K. It appears white because.
   A. Most of the light emitted is white light.
   B. Most of the light emitted is red light.
   C. Most of the light emitted is blue light.
   D. Most of the light emitted is yellow and green.

2. Sending more electrical current through a light bulb will make the filament hotter. How does the color of the light bulb change as the current is increased?
   A. The bulb gets brighter making it only appear to change color.
   B. The color shifts to a shorter wavelength.
   C. The color shifts to a longer wavelength.
   D. The color change is an optical illusion.

3. A 40 W fluorescent bulb gives off much more light than a 40 W incandescent bulb. Why is this true?
   A. Fluorescent bulbs give off more ultraviolet light.
   B. Incandescent bulbs give off more infrared light.
   C. Incandescent bulbs use more energy per second.
   D. Fluorescent bulbs use more energy per second.

4. Einsteinium, $^{254}_{99}Es$, undergoes alpha decay. If the alpha particle is emitted with a speed of 2×10^6 m/s, find the recoil speed of the daughter nucleus? What is the daughter nucleus?

5. A radioactive sample is emitting both alpha and beta rays. If an external magnetic field is present, how will you be able to distinguish the two?
   A. The alpha rays will bend in one direction and the beta rays the other.
   B. The alpha rays will bend, but the beta rays will not.
   C. The beta rays will bend, but the alpha rays will not.
   D. Neither rays will bend.

6. Lead, $^{214}_{82}Pb$, decays in several steps. Starting with the lead, the nucleus emits 2 betas, 1 alpha, 1 beta, 1 alpha, and 1 beta in succession. What is the final daughter product?

7. In the Sun, a series of nuclear reactions have the net effect of making 1 helium atom from 4 hydrogen atoms. This is an example of:
   A. nuclear fission
   B. nuclear fusion
   C. nuclear reactor
   D. nuclear chain reaction

8. Potassium-40 ($^{40}_{19}$K) is radioactive and decays by electron capture. The daughter nucleus of this electron capture is:
   A. Potassium
   B. Argon
   C. Calcium
   D. Chlorine

9. A series of radioactive decays begins with uranium, $^{234}_{92}$U. The decay steps in the series are: 4 alphas, 1 beta, 2 alphas, 3 betas, 1 alpha. The final daughter product is:
   A. $^{206}_{82}$Pb
   B. $^{202}_{82}$Pb
   C. $^{210}_{82}$Pb
   D. $^{206}_{84}$Po

10. In the Sun, a series of nuclear reactions have the net effect of making 1 helium atom from 4 hydrogen atoms. The energy released for each helium atom produced is about 25 MeV. (1 u = 1.66 x 10^{-27} kg = 931.5 Mev/c^2) The difference in mass between 4 hydrogen atoms and 1 helium atom is:
   A. one helium is the same mass as 4 hydrogens.
   B. helium is 25 MeV lighter.
   C. helium is 0.027 u heavier.
   D. helium is 0.027 u lighter.

11. A certain sample of radioactive material has a half-life of 10 hours. After 30 hours, the activity of the sample is 80 millicuries. A curie is a measure of the number of decays per second. The initial activity of the sample must have been:
   A. 10 millicuries
   B. 20 millicuries
   C. 320 millicuries
   D. 640 millicuries

12. The amount of charge lost by a charged capacitor is proportional to the time and the charge on the capacitor. A certain capacitor is charged up to a voltage of 40 volts and then observed to drop to 20 volts after 3 seconds. How much voltage will be across the capacitor after another 6 seconds?

Answers to Questions and Problems

1. White light is what our brain says we see when our eyes are receiving light of all colors, but the energy peaks in the yellow-green region of the spectrum. The answer is D.

2. Wien's law says that at the temperature increases, the peak wavelength decreases. The answer is B. The bulb gets brighter (A), but the color also changes. Make sure you read all the answers before choosing one.

3. The 40 W means that each bulb uses electrical energy at the rate of 40 W. Both use the same energy per second. Incandescent bulbs waste more of this electrical energy than fluorescent bulbs. The temperature of the filament of an incandescent bulb must be relatively low or the filament would melt. At this temperature the filament emits most of its energy as infrared light (which we feel as "heat" as our body absorbs it). Fluorescent tubes contain mercury vapor which emits a bright line spectrum and is more efficient at converting electrical energy to visible light energy. The answer is B.

4. The daughter nucleus has lost 2 protons (and 2 neutrons) so it is element number 97 which is berkelium. It has an atomic mass of 250 since the einsteinium lost 4 nucleons. We use conservation of momentum to find the speed of the berkelium, $^{250}_{97}Bk$, nucleus. The momentum of the alpha particle in one direction is equal to the momentum of the berkelium nucleus in the other direction. $MV = mv$ (The atomic masses of the particles are approximately equal to the atomic mass number.)
    $(250)V = (4)(2 \times 10^6 \text{ m/s})$
    $V = (4)(2 \times 10^6 \text{ m/s})/(250) = 3.2 \times 10^4 \text{ m/s}$

5. Since alpha particles are positively charged and beta particles are negatively charged (assuming that the question is referring to emitted electrons), they will bend in opposite directions. The only possible answer is A. Positrons (β^+) would bend in the same direction as alpha rays.

6. Notice that the emission of 1 alpha and 2 betas gets you back to the same element, but with 4 less neutrons. (The betas are assumed to be electrons.) Since a total of 2 alphas and 4 betas were emitted in the series, the final daughter must be the same element we started with. The final daughter is lead, $^{206}_{82}Pb$.

7. Since 4 hydrogens are fused together to make one helium, this is an example of nuclear fusion. The answer is B.

8. The captured electron converts a proton to a neutron. The atomic number is *reduced* by one (to 18). The atomic mass number does not change, so the daughter nucleus is argon-40. The answer is B.

9. A total of 7 alphas are emitted which reduce the mass number by:
4 x 7 = 28 Therefore, the mass number of the daughter is 206. The 7 alphas *decrease* the atomic number by 2 x 7 = 14, and the 4 betas *increase* the atomic number by 4, resulting in a net decrease of 10. The atomic number of the daughter is 82 so the answer is A.

10. Since energy is emitted by the Sun and the nuclear reactions that fuel the Sun, one helium must be less massive than 4 hydrogen. The mass difference is: (1 u)(25 MeV)/(931.5 MeV) = 0.027 u The answer is D Regarding answer A, helium is 25 MeV/c^2 lighter, not 25 MeV lighter.

11. Thirty hours is 3 half-lives. The activity of the sample must have decreased by 1/2 for each half life. Only C and D are possible answers. Going backwards in time, the activity would increase each half-life. Three doublings of 80 gives 640. The answer is D.

12. The capacitor discharges exponentially since the charge lost is proportional to the time and charge on the capacitor. (These are the conditions that radioactive elements satisfy, leading to their exponential decay.) Therefore, the "half-life" of this capacitor is 3 seconds. After 6 seconds (2 more half-lives) the capacitor's voltage will drop by a factor of (1/2) = 1/4. The 20 volts will drop to 5 volts after another 6 seconds.